EAI/Springer Innovations in Communication and Computing

Series Editor

Imrich Chlamtac, European Alliance for Innovation, Ghent, Belgium

AF802157

Editor's Note

The impact of information technologies is creating a new world yet not fully understood. The extent and speed of economic, life style and social changes already perceived in everyday life is hard to estimate without understanding the technological driving forces behind it. This series presents contributed volumes featuring the latest research and development in the various information engineering technologies that play a key role in this process.

The range of topics, focusing primarily on communications and computing engineering include, but are not limited to, wireless networks; mobile communication; design and learning; gaming; interaction; e-health and pervasive healthcare; energy management; smart grids; internet of things; cognitive radio networks; computation; cloud computing; ubiquitous connectivity, and in mode general smart living, smart cities, Internet of Things and more. The series publishes a combination of expanded papers selected from hosted and sponsored European Alliance for Innovation (EAI) conferences that present cutting edge, global research as well as provide new perspectives on traditional related engineering fields. This content, complemented with open calls for contribution of book titles and individual chapters, together maintain Springer's and EAI's high standards of academic excellence. The audience for the books consists of researchers, industry professionals, advanced level students as well as practitioners in related fields of activity include information and communication specialists, security experts, economists, urban planners, doctors, and in general representatives in all those walks of life affected ad contributing to the information revolution.

About EAI

EAI is a grassroots member organization initiated through cooperation between businesses, public, private and government organizations to address the global challenges of Europe's future competitiveness and link the European Research community with its counterparts around the globe. EAI reaches out to hundreds of thousands of individual subscribers on all continents and collaborates with an institutional member base including Fortune 500 companies, government organizations, and educational institutions, provide a free research and innovation platform.

Through its open free membership model EAI promotes a new research and innovation culture based on collaboration, connectivity and recognition of excellence by community.

More information about this series at http://www.springer.com/series/15427

G. R. Kanagachidambaresan • R. Anand
E. Balasubramanian • V. Mahima
Editors

Internet of Things for Industry 4.0

Design, Challenges and Solutions

Springer

EAI
RESEARCH MEETS INNOVATION

Editors
G. R. Kanagachidambaresan
Department of CSE
Vel Tech Rangarajan Dr Sagunthala R&D
Institute of Science and Technology
Chennai, India

E. Balasubramanian
Department of Mechanical Engineering
Vel Tech Rangarajan Dr Sagunthala R&D
Institute of Science and Technology
Chennai, India

R. Anand
Department of EEE
Amrita Vishwa Vidyapeetham University
Bangalore, India

V. Mahima
Department of ECE
Vel Tech Rangarajan Dr Sagunthala R&D
Institute of Science and Technology
Chennai, India

ISSN 2522-8595 ISSN 2522-8609 (electronic)
EAI/Springer Innovations in Communication and Computing
ISBN 978-3-030-32532-9 ISBN 978-3-030-32530-5 (eBook)
https://doi.org/10.1007/978-3-030-32530-5

© Springer Nature Switzerland AG 2020
This work is subject to copyright. All rights are reserved by the Publisher, whether the whole or part of the material is concerned, specifically the rights of translation, reprinting, reuse of illustrations, recitation, broadcasting, reproduction on microfilms or in any other physical way, and transmission or information storage and retrieval, electronic adaptation, computer software, or by similar or dissimilar methodology now known or hereafter developed.
The use of general descriptive names, registered names, trademarks, service marks, etc. in this publication does not imply, even in the absence of a specific statement, that such names are exempt from the relevant protective laws and regulations and therefore free for general use.
The publisher, the authors, and the editors are safe to assume that the advice and information in this book are believed to be true and accurate at the date of publication. Neither the publisher nor the authors or the editors give a warranty, express or implied, with respect to the material contained herein or for any errors or omissions that may have been made. The publisher remains neutral with regard to jurisdictional claims in published maps and institutional affiliations.

This Springer imprint is published by the registered company Springer Nature Switzerland AG
The registered company address is: Gewerbestrasse 11, 6330 Cham, Switzerland

To our students, parents, and the almighty

Preface

Internet of Things (IoT) has become the inevitable component in smart city development. The decline in the workforce of developing countries is presently solved through cyber-physical system (CPS) technology. The industries are enabled with CPS to monitor their production, enhancing their productivity through Industry 4.0 standards. The main objective of Industry 4.0 is to integrate all the machines as a single integrated network for M2M communication. The CPS is enabled in autonomous robots, UAVs, cybersecurity, cloud storage systems, additive manufacturing, augmented realities, and big data computations. The production became noticeable during Industry 3.0 standards; however, the microcomputer interface and machine-to-machine interaction in Industry 4.0 have made a tremendous increase in production. Industry 4.0 has decentralized analytics, critical decision making, and increase in response time during productions. The machine-to-machine interaction has become possible through wireless and wired communication. Robust and foolproof system design is required to ensure workers safety. The existing companies with 3.0 standards are facing a formidable challenge to meet 4.0 standards. The existing CNC machines cooperation with CPS is tedious, and researchers are working toward fault-tolerant algorithm design for the same. Creating transparency in production between joint ventures has brought greater impact on profit and production. An efficient venture makes the system prone to cyber-based attacks. Researchers are focusing on lightweight cryptography algorithm to prevent the CPS from attacks and provide high security. These connected machines in industries provide a high volume of data, informing the nature of the machine and its present state of working. The machine about to fail can easily be predicted through prognostics and artificial intelligence. The system has the capability to identify the broken parts and their faults to the user. Deep learning and machine learning algorithms can ensure workers safety and can avoid industrial accidents. The role of robotics and additive manufacturing through 3D printing has attracted a lot of young researchers in realizing their ideas through rapid prototyping. These chapters provide an insight for the development of algorithms in Industrial 4.0 scenario. The book concentrates on providing an overview in areas such as robotics, RFID, image processing, workers safety, software-defined network, and smart grid applications.

Chennai, Tamil Nadu, India G. R. Kanagachidambaresan
Bangalore, Karnataka, India R. Anand
Chennai, Tamil Nadu, India E. Balasubramanian
Chennai, Tamil Nadu, India V. Mahima

Acknowledgment

We would like to thank Imrich the Series Editor and Eliska the Managing Editor at EAI for giving us the opportunity to edit a book in Innovations in Communication and Computing (Springer). We thank the authors who have put tremendous efforts in this book project. We would also like to thank the reviewers who give an unbiased review on time for successful completion. The generous support from Vel Tech Rangarajan Dr. Sagunthala R&D Institute of Science and Technology, Amrita Vishwa Vidyapeetham, Bangalore, allowed contributing an exciting interesting material to new researchers and students. We hope this book would be a necessary material for engineers and scientists.

Contents

About the Editors

G.R. Kanagachidambaresan had completed his PhD in Information and Communication Engineering at Anna University, Chennai. He has successfully completed numerous consultancy projects and obtained patents. He has authored many research articles in leading journals and conferences. He is serving as an editorial board member in many leading journals. He is presently working in projects funded from Indian Space Research Organisation (ISRO). He has edited several books and also guest edited many special issues. He is a software solution provider for companies and industries. He holds numerous copyrights and patents on software developed and product designed. He is presently working on Internet of Things applications and remote health monitoring.

R. Anand completed his PhD in EEE at Anna University, Chennai. He has successfully authored articles in leading journals and conferences. He has also carried out many consultancy works and projects for MNCs. He is presently an assistant professor in the Department of EEE, Amrita Vishwa Vidyapeetham University, Bangalore, India. He has successfully hosted many special issues in leading journals.

E. Balasubramanian had completed PhD in the field of Robotics and Control at Concordia University, Montreal, Canada. He has more than 15 years of experience in carrying out numerous research projects. He executed consultancy projects for national agencies such as BARC, IGCAR, BEL, ISRO, and automobile sector. He collaborated with Taiwan scientists under Indo-Taiwan schema for the development of Micro Aerial Vehicles and SWARM control through image processing techniques. He is presently working on Indo-Canada collaborative research project aiming at inspection of railway bridges and heritage structures using UAVs and also cooperating with Korean scientists (Indo-Korea Collaboration) to develop unmanned amphibious vehicle to collect water samples for water quality monitoring. His team developed unmanned aerial vehicles (UAVs) for diverse applications including power line inspection, telecom tower inspection and radiation measurement, environmental monitoring, surveillance, and aerial view inspection. His team won the first prize of INR 5 lakhs in a national-level competition organized by Power Grid Corporation Limited and the prestigious telecom sector Aegis Graham Bell award in the category of best innovative business model 2014. His research interests are UAVs, robotics, and control and sensors with online data acquisition systems.

V. Mahima is presently an assistant professor in the Department of ECE in Vel Tech Rangarajan Dr Sagunthala R&D Institute of Science and Technology, Avadi, Chennai. She has successfully authored many articles and projects. Her areas of interest include Wireless Sensor Network, Wireless Body Sensor Network, Internet of Things, and Industry 4.0.

Chapter 1
A Survey on RFID in Industry 4.0

Ermal Elbasani, Pattamaset Siriporn, and Jae Sung Choi

Introduction

The advanced and rapid achievements of information and communication technology have brought in a new industrial revolution, so-called Industry 4.0. This has come with new challenges in science and industry, but specific efforts need to be done to demonstrate that new industrial systems can be applied to factory environments. Mentioning important topics of Industry 4.0 is RFID system, and Internet of Things (IoT) technology, with the function to generate data from target objects and then access an Internet. Moreover, RFID system can be a role as a link between process flow data and physical asset data. IoT assists accumulate a lot of data through RFID tags and readers and sensors. And it improves transparency and efficiency of production management with big data analysis and artificial intelligence technologies.

Despite RFID as main part of IoT, in Industry 4.0, Internet of Services (IoS) and cyber-physical systems (CPS) play a key role in production process. Also, much research is being done on RFID applications in industry such as inventory management system [1], object location detection [2], RFID using MES [3], planning and scheduling decision support system [4–6], material flow monitoring system [7, 8], and quality assurance system [9].

This research aims to review the Industry 4.0 studies on the viewpoint of RFID, since on our good knowledge does not exist any similar review with particular focus on the significance of RFID in the Industry 4.0.

E. Elbasani · P. Siriporn · J. S. Choi (✉)
Department of Computer Engineering, Sun Moon University, Asan, South Korea
e-mail: jschoi@sunmoon.ac.kr

© Springer Nature Switzerland AG 2020

G. R. Kanagachidambaresan et al. (eds.), *Internet of Things for Industry 4.0*,
EAI/Springer Innovations in Communication and Computing,
https://doi.org/10.1007/978-3-030-32530-5_1

History of RFID in Industry 4.0

RFID in History

RFID technology has evolved rapidly over years, particularly widely applied with the advancements of web technology. The initial appearance of RFID technology was in World War II, used in a British project that developed a system to detect and identify whether fighter aircraft belong to enemy or their ally, using RF technology.

And the fourth industrial revolution is when through Internet or the web technology and the physical industrial technologies combined into one. Since the early 1990s, Internet has had revolutionary impact on living, culture, and commerce. In fourth industrial revolution, RFID can be the main source of data generation in smart environments, such as smart factory, with utilization of Internet technologies.

RFID Technology

RFID system itself is simple. RFID is found in long-range and low-range format. Long-range RFID referred also as RAIN RFID, based on the UHF band, has the fastest recognition speed and can read tags up to 15 m, and it is suitable for factory usage.

The low-range RFID uses the HF band, provides low tag recognition, and reads till 1.2 m. The low-range RFID reader is suitable for smartphone users and has found a great usage in consumer-trade shopping activities and also other areas like transport, etc. With the development of technology, there are being invented a lot of ideas of using it.

In manufacturing, RFID can be an important tool to help manufacturers efficiently produce more products, and also machines can communicate with each other and perform an immediate action if a mistake occurs.

RFID systems are ideal for any company that is required or wants to utilize automated processes and interconnect modules and devices. Industry 4.0 has the opportunity to be easily embraced in the manufacturing industry, and RFID has the ability to realize everything. Passive RFID has been used massively in the past years, and it is easily expected the passive RFID system will be continuously and widely applied in our industries, especially in the ultrahigh-frequency range, and also active RFID not as widely, but with the advantage for real-time location system (RTLS) applications [10].

In order to make active RFID usefully and popularly the same, it is required standardization of active 433 MHz RFID to drive growth. Also, mentioning, ultra-wideband (UWB) and Bluetooth low-energy (BLE) RTLS solutions will further enhance the solution [11].

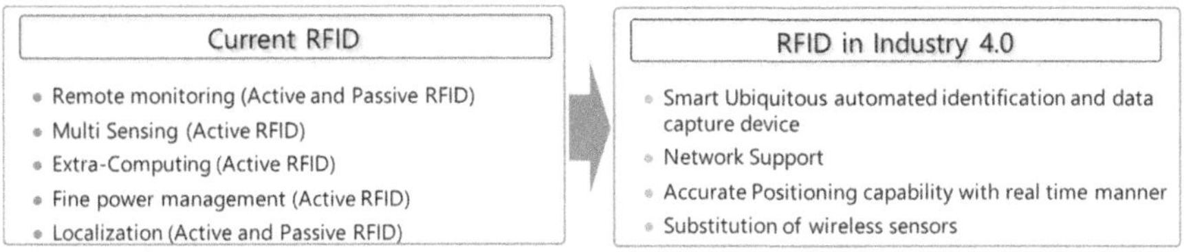

Fig. 1.1 Development of the RFID toward the industrial IoT and Industry 4.0

A significant concept of RFID is generally shown in Fig. 1.1, that describe how the main components of RFID develop to be embedded in smart Industrial technology. Over the past decade, remote monitoring, used by RFID and IoT to manage multiple sensor inputs and using RFID in localization, has been an attractive issue. Industry 4.0 is moving to focus on smart products, analyze RFID/IoT data, and research on network functions to improve system performance.

RFID and Big Data

Nowadays, many factories have begun involving advanced information technologies and knowledge discovering methods to facilitate information flow, with the main need to deal with the large volume of data which are generated and collected from manufacturing processes and its related operations. The generated and collected data from processes in factory such as production, transporting, surveillance, etc. is growing exponentially [12].

Particularly in Industry 4.0, data integration and analyzing is applied to improve in decision-making stage, increasing performance in scalable data flow adaptation processes and systems [13]. Big data issue deals with large volume that need to be processed in time as quick as possible and with the ability to combine various data formats.

One another challenge of passive RFID tags is to adapt to complex, dynamic manufacturing environments since they generate relatively small storage, compared to large number of real-time data record processes. Also for a high decision-making standard, it requires a universal and effective network connection to arrange data on the network. This challenges the data transferring and network bandwidth when dealing large number of RFID tags emitting signals to be recorded and analyzed. Active RFID tags, which occupy large storage space, make able for higher intercommunication among machines, and also workers have more ease to detect and cope with different events and make the proper decision.

The big data technologies are a powerful and useful way to address variety and huge-capacity data. Next there are explained in sequentially five steps the concept of RFDI data processing from signal capturing to decision-making stage, also illustrated in Fig. 1.2.

Fig. 1.2 RFID data processing phase

First is collection phase: it has to do with capturing data signals from RFID readers and RFID tags and also from smart sensors distributed. RFID raw data is stored in a database in tables format. In this stage it is challenging to make the most of knowledge to support advanced decision.

Second is data cleaning: RFID devices and IoT devices generate large-amount raw data. These data are automatically correlated with process flow. However, due to the environment effect, the random nature of RFID events, there may be many incomplete, and inaccurate, data. And in this stage, the data goes in a process of organizing based on some programmed standards from the upper-level stages.

The third step is data integration. In the case of massive data, systems need to configure real-time and non-real-time data and structured and unstructured data. The integration process, which adopted data mining methods, is by clustering, classification, association, and prediction using decision trees, genetic algorithms, and supported vector machines techniques.

This step helps you re-configure valid data according to process flow order and time sequence to translate valid data into meaningful pattern information and knowledge. Inputs of this stage are cleaned and organized data, and outputs are data that flows according to process order.

The fourth step consists of knowledge extraction. The purpose of this step is to transform the organized data into meaningful information to support control and decision-making for output control. Inputs are organized data according to process, and outputs are information derived from useful attributes and data features. SVM (support vector machine) [14], k-means clustering algorithm, and deep neural network (DNN) can be used to process large RFID-based production data. You can discover knowledge, patterns, and rules based on the processing results.

The final stage of decision-making and influential work is to interpret the information, knowledge, patterns, and rules achieved in a useful and simple way.

When IoT and RFID are applied to the manufacturing process, heterogeneous RFID-based data, real time and in huge amount data, are generated and recorded, and that is called industrial big data [15, 16].

In [17, 18] researches provide a broad overview and studies about variety of management of captured data by RFID and related mining, analyzing, and processing techniques. The large amounts of data generated from sensing devices, mobile devices, and RFID devices are termed as big data.

Mining techniques for RFID data help to find a pattern regarded to process when it is applied. This means that this pattern matches the respective frequent process, with the purpose to provide concrete knowledge to user/owner and assure also a visual information [19].

In the analyzing process, RFID data mining algorithms, which are utilized with SVM (support vector machines), DT (decision tree), and NN (neural networks) [14], have been adopted for cases of clustering, classification, and deep learning. In addition, the mining method is used in operational research methods, combining mixed integer and probability-reduction programming. The most highlighted methods of data mining are explained as follows:

- Associative Rules: In mining methods in data science, associative rules are one of the most studied methods. In [11], associative rules include both positive and negative cases. For instance, a positive association is when spatiotemporal data are shown together, and where the target is not displayed in the location is a negative association. However, as the volume of RFID data increases, the current associative methods require some modification in order to gather more useful data. In [20] is explained a combined ontology-based rules with hierarchical clustering associative rules for reducing duplication of data during RFID data mining process. In addition, [9] describe a new approach of multiple associative rules to reduce RFID data abstraction level. The advantages and disadvantages of the given algorithm are that it provides a new research direction for RFID path data mining.
- Cluster Analysis: Clustering algorithm is also mainly used for RFID data mining. It creates groups of data with high similarity or for various features of data [11]. The primary purpose of this rule is to reduce repetition and uselessly categorize the gathered path data into the DB. While the data tags are clustered, every manufacturing work is associated with data about space, time, and instruction of operation. Thus, the flow of RFID data can be related to statistics or machine learning model for analysis [21].
- Frequent Path Data Mining: Since path tracking and tracing are always useful information to management and control center, RFID has been used to provide an effect with relatively cheap cost to monitor location and position. Furthermore, its efficiency is based on mining tools for pattern recognition [22]. Many algorithms have been introduced over the period [23], since problem formulas and path data mining algorithms vary depending on the application situation. As the usefulness of RFID technology increases, frequent path mining methods will draw more attention specifically in industry.

Data includes useful information with great value or also useful pattern recognition solutions. Generating efficient information from the RFID data can help with task management, planning, and error detection, position and localization, and tracking and tracing.

RFID in Internet of Things

About RFID with IoT

Internet of Things was briefly described as an integration of network which collects and exchanged data acquired through embedded sensors, actuators, or other devices in things or physical objects in the world [24, 25].

The evolution of IoT over the three generations, as shown in Fig. 1.3, is characterized by enabling key technologies, major enhancing of architectural solutions and available products [26]. The objectives founding IoT in the first generation are to identify and combine data from objects (tagged things) using the capabilities of RFID and sensor connected through wireless sensor network systems (WSNs), which have led the use of the Electronic Product Code (EPC) and machine-to-machine communication (M2M) in industry. The objectives of the second generation, the full interconnection of things and Web of Things, focus on directly connecting the objects to the Internet using technologies for integrating devices into IP network and Web of Thing technologies. The third IoT generation is the era of Social Internet of Things, cloud computing, information-centric networking for IoT. This generation focuses on making intelligent objects community, exploiting cloud computing technologies for storage, communications and processing capabilities, and shifting from host-centric to content-centric networking.

Different IoT architectures have been proposed by different researchers. For example, Atzori et al. [27] proposed to develop an architecture with three layers for IoT such as the sensing, the network, and the application layers. Jia et al. [28] divided the architecture into perception, network, and service or application layers. Liu et al. [29] proposed to design an IoT application infrastructure that consists of physical, transport, middleware, and application layers. Yaqoob et al. [30] divided IoT architecture into four layers of sensors and connectivity, gateway and

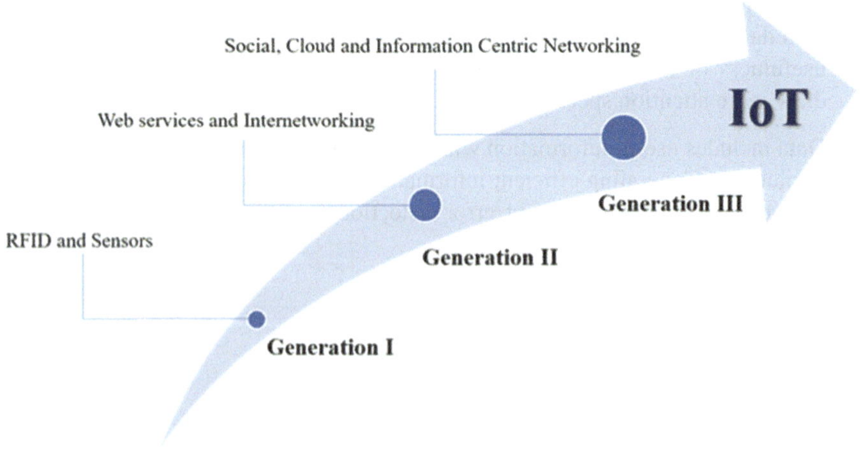

Fig. 1.3 Evolution of IoT

network, management and security services, and smart applications. Da et al. [31] described the four-layer architectures for service-oriented IoT that contains sensing layer, networking layer, service layer, and interface layer, and the whole layers of architecture must enforce security. In specific for industrial IoT applications, the design architecture needs to consider energy, latency, throughput, scalability, topology, security, and safety. Mashal et al. [32] proposed five layers of IoT architecture which contains perception layer, network layer, middleware layer, application layer, and business layer and that architecture is added two new layers of the processing layer and business layer because three-layer architecture is not enough due to the development in IoT. Kumar and Mallick [33] summarized the different IoT architecture layers as three-layer, five-layer, modified five-layer, six-layer, seven-layer, and FOG architecture. The modified five-layer architecture consists of application layer, sensing layer, communication layer, service layer, and infrastructure layer. The six-layer architecture comprises focus layer, cognizance layer, transmission layer, infrastructure layer, and competence business layer. The seven-layer composes of application layer, ASM layer, service layer, communication layer, network layer, hardware layer, and environment. The FOG computing layer, where smart sensors and smart network gateways do a part of the data processing and analytics, consists of physical layer, monitoring layer, preprocessing layer, storage layer, security layer, and transport layer, and that inserts monitoring, preprocessing, storage, and security layers between the physical and transport layers.

Furthermore, industry has utilized the advantage of IoT to enhance industrial systems and applications. This rises to the industrial IoT (IIoT) and enables to interconnect humans, machines, and devices. In order to apply IoT in the industry, it needs to consider technology requirement and architecture design needed for numerous devices coordination system as trillion connected devices in the future of the IoT system.

The Industrial Internet Consortium (IIC) [34] has provided Industrial Internet Reference Architecture (IIRA) with the concept of industrial Internet system (IIS). The IIRA drives IIoT systems through Industrial Internet Architecture Framework (IIAF) based on ISO/IEC/IEEE 42010:2011.

For Industry 4.0 as subset of IIoT, the Working Group for Industry 4.0 [35] have provided the Reference Architecture Model for Industry 4.0 (RAMI4.0), which consists of a three-dimensional model and six layers of IT to representation of Industry 4.0.

Industry has been taking advantage of IoT which is tending to be pervasively used in various industries. The foundational technology of the IoT, such as RFID, has been widely used in industry since 1980. RFID, which refers to radio-frequency identification, provides the contributions of identifying and tracking objects to gather data. Electronic Product Code (EPC) was developed by Auto-ID Centre at MIT in 1999. EPC is electronic code stored data from RFID and serves as a universal RFID system. The use of RFID for sharing information has EPCglobal as standardization of EPC. According to the previous described architectures of IoT, RFID and IoT can incorporate to embed RFID and gather information from RFID-enabled physical objects.

IoT makes objects sense physical real world through Internet protocol and wire or wireless networks connecting and embedded sensors such as RFID, which can be embedded into the objects and acquires data from "things" or physical objects. IoT and RFID can be applied to production processes in order to control and improve productivity in manufacturing.

Additionally, one of the technologies in automated identification and data capture (AIDC) is RFID, which can be used to automatically identify objects and directly access data processing systems. Integrating RFID and IoT will provide for monitoring objects embedded tags.

RFID technology is not only enabled to embed on physical objects with purpose to gathering the physical information but also enabled to monitoring of chemicals, gas, temperature, and humidity, during production or shipment. RFID was implemented in the form of interdigitated chemo-capacitance provided with humidity-sensitive polymers. Oprea, A. et al. presented the development of novel polymer-based capacitive humidity sensors integrated onto flexible RFID label substrates [36]. In the perishables supply chain, RFID-based sensors play the important role to detect the quality of perishables goods and examine the condition of perishable product during shipment through RFID-based temperature sensors [37].

Localization is one of RFID applications that has been used with the benefit of passive RFID tags, no battery and cheap cost. RFID localization can be categorized by different techniques as tag and reader localizations. For tag localization technique, an object or human needs to carry the tag, which will be detected through multi-reader in environment, while for reader localization technique, the single reader will read signal of multi-tags deployed in the environment instead of carrying the tag on the object or human.

Additionally, Logistics 4.0 can use RFID technologies for resource planning, warehouse management systems, transportation management systems, intelligent transportation systems, and information security [38].

RFID-Based IoT in Industry 4.0

One of the core technologies, which have been used in Fourth Industry Revolution, is the Internet of Things. That is the reason why Industry 4.0 is also known as industrial Internet. IoT is able to connect physical objects, systems, and services, enable object-to-object communication, and share physical information objects [39]. Among IoT technologies, RFID is widely used as a cheap cost with powerful capability to identify, trace, and track physical objects, providing real-time data about the relative devices, decreasing labor, simplifying process, increasing the accuracy about storages, and improving efficiency [31]. The integration of IoT and RFID into Industry 4.0 enables a supply chain management, smart factory, as well as a smart city. RFID was used to identify the good and monitor the stocks in warehouse management.

Supply Chain Management

Supply chain management runs through among supply chain partners supplier, manufacturer, distributor, retailer, and consumer. The integration of IoT in the supply chain provides performance in every single of the processes in manufacturing environments; man-to-man, man-to-machine, and machine-to-machine connections are provided for intelligent perception [40]. IoT-enabled manufacturing characterizes real-time data gathering and sharing among various devices, operators, and other manufacturing resources [41].

IoT technologies help to monitor industrial operations and provide supply chain visibility and traceability for the movements of resource and related information flows in manufacturing operations [4, 42]. Supply chain visibility is to track and trace a product from origin to destination along the logistics promises to enable real-time tracking of material flows, improved transport handling, as well as accurate risk management, lay of the supply chain. It means all of the plays in supply chain are able to be linked to each together and easily achieve procedures from supplier to customer service. The important need is communication and exchanging information with each other.

The technology of RFID can give information of physical objects and data collected on several points in the supply chain with longer range of read better than bar code technology and easier use for manufacturer. The integration of RFID and IoT aims to be efficient sharing of resources controlling quality can be enabled by the application of RFID in manufacturing shown in [7, 43–47].

The goal of logistic is not only transportation products in the right place, at the right time, but also in the right quality. Thus, the term "Logistics 4.0" has been highlighted in the past years. Also, Logistics 4.0 is concerned to the same conditions as smart services and products, which are used to define "Smart Logistics" [48].

RFID is capable of identifying and capturing shipping products [49] described logistics-oriented Industry 4.0 application model in two dimensions as physical supply chain dimension and digital data value chain dimension. The physical supply chain dimension is a level of automatic interaction along with physical thing; RFID and IoT system can be used among two dimensions, for example, gathering data, tracking and tracing, and making services along the supply chain automatically.

In 2003, the world's leading retailers, Metro Group in Germany, Marks & Spencer in the UK, and Walmart in the USA, utilize RFID in tracking supplies [50]. Fashion retailers also use RFID in supplies [51, 52].

Passive RFID tags can be exploited to detect and record customer's activity or shopping behaviors. For example, ShopMiner, RFID-Based Customer Shopping Behavior Mining System, is able to identify customer shopping behaviors with high accuracy and low overhead and is robust to interference [53]. RFID is not only able to collect customer shopping behavior but also enhance customer shopping experience [54]. The requested services of a customer need to be trustable and make the customer use the service and share their data without any concerns [26].

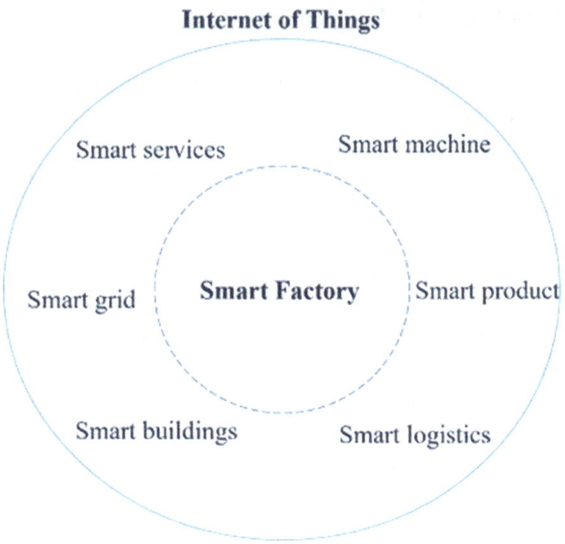

Fig. 1.4 Smart factory

Smart Factory

Industry 4.0 is the link of IoT and the digital supply chain partners around a smart factory that build up the business. A general concept of the main contributors is shown in Fig. 1.4. Each digital supply chain links applications to the business process and each other as well. In addition, the smart factory relies on smart manufacturing. The smart manufacturing connects all machines, sensors, or devices in the manufacturing to intercorporate each other. The goal of smart manufacturing is to run production and product transactions using advanced information and manufacturing technologies smoothly and consistently. The approaches of smart manufacturing, such as cyber-physical systems (CPS) and IoT frameworks can be used for data collected and analyzed in real time through IoT [55].

The smart factory connects physical and digital world based on CPS. Wang et al. [56] presented a conceptual design of smart factory framework consisting of four layers, namely, physical resource layer, industrial network layer, cloud layer, and terminals for supervision and control layer. The smart things may communicate through the industry network to collect physical resource information in the cloud and allow human to interact through the terminals. The smart factory leads to optimize production in order to improve flexibility, productivity, resource and energy efficiency, transparency, promoting integration, profitability, and friendliness to labor. RFID can develop along with the smart factory through the advantages of RFID tags. By using RFID tags embedded on the products, the raw products in smart factory processes will be read and written so as to update information during production.

RFID in Communication and Networking for Industry 4.0

Industry 4.0 uses both a smart object and network for the integration of physical devices into the network during producing and manufacturing processes. Communications and networking can be evaluated as a comprehensive in-depth technology for capabilities of Industry 4.0 features, such as analytics of large data, embedded systems, and simulations also. RFID and RTLS (Real-Time Location Tracking System) are one of the most supported technologies in Industry 4.0. Systems based on RFID and RTLS are achieving: [57]. Process-optimized product production with a large number of versions, improved function and flexible assembly, and high level of data transparency.

Mobile technology has shown significant strides since they first came out, and now they are more than just communication devices. These tools allow the Internet to receive and process large amounts of data and provide in-built sensors such as high-quality cameras and microphones to record and transmit the recorded. Mobile service is a feature provided by mobile device software. The software uses network, integrated hardware, sensors, storage, and multimedia applications to deliver these capabilities. Most devices (smart devices) these days come with some of the manufacturer's own built-in services. Other services can be provided by the network operator. Smartphone devices can be integrated with third-party service providers or system integrators to enable connectivity on the network.

Specific case of mobile RFID method is when mobile services use RFID as a technology to access information on objects tagged with RFID tag [58]. The mobile RFID service simply provides information about objects tagged with RFID. Mobile RFID services use services in relation with RFID reader and back-end network infrastructure, but network and back-end servers are not always required to deliver services. When it comes to mobile RFID, it's a different technique from traditional RFID. In mobile RFID, the tag is fixed and the reader is mobile and not the other way around. There are certain advantages over traditional RFID. It's enough if it is not required a wire connected and only needs a mobile reader that can cover a wide area instead of several of fixed readers.

Also, people are always carrying cell phones. Smart mobiles are devices with a sense of intelligence, embedded with a lot of tiny sensors [59]. They can see, hear, identify, track, and measure perfectly. People frequently are in mobile state and continuously in need to interact with their surroundings every day. In an environment when objects or devices become smarter, by utilizing various sensors and communication facilities, will bring in creating a smarter environment around people.

Mobile RFID services needed a mobile RFID platform that can be easily delivered using smart mobile devices. Cell phones integrated with RFID readers are called mobile RFID readers. It is able to create a hybrid reader by combining UHF readers with other identification technologies such as NFC and barcode. This combination creates a unique identification platform. Smartphones with an RFID-integrated platform will be a key driver of the IOT service. Ubiquitous services are available in a ubiquitous environment. Mobile RFID services are one driving force to achieve these goals.

Conclusions

Roles of RFID Technology in Retail

RFID technology in the supply chain is to transfer information in digital format and then collect data from a variety of objects available to everyone. RFID technology also has a great advantage in tracking and positioning individuals and objects in factories or in other locations within the supply chain.

Meanwhile collecting massive amounts of data from enterprise operating systems can be a serious factor. Reading multiple tags at the same time can cause conflicts and ultimately result in data loss. Therefore, studies are continuing on new ways to prevent this phenomenon while anti-collision algorithms are applied. RFID can be useful such as operational efficiency, enhanced visibility, reduced costs, increased security, improved customer service, improved information accuracy, and increased revenue. Table 1.1 lists the main tasks that include RFID that is most relevant to real industry.

Roles of RFID Technology in Logistics

RFID technology has shown great benefits of noncontact, non-line-of-sight, automatic identification and convenient operation, making it ideal for material tracking and vehicle and shelf identification. The application of RFID technology effectively implements logistics management automation and intelligence. This automation and intelligence is primarily implemented in logistics operations management and supply chain management.

RFID in the logistics operations management application is primarily performed in transportation and distribution management, warehouse storage and material

Table 1.1 Advantages and disadvantages of RFID applications

Application	Advantages	Disadvantages
Retail	Localization	Tag collisions
	Preventing the return of expired products in the supply chain	Privacy for consumer can be violated
	Unique ID	Metal-liquid signal absorption
		Data integrations
Logistic	Identification of goods	Security
	Improve the quality of inventory operations	Tag collision
	Improve the meticulous management	Multi-RFID reader needed
	Anti-counterfeit	
	Enhance the scope of information sharing	
Manufacturing	Streamline asset tracking	Data management
	Tracking with higher accuracy	*Not all companies use RFID*

management, and distribution processing management. RFID optimizes operations management by quickly and accurately tracking each homework link in real time through the entire logistics operations management process [60, 61].

Roles of RFID Technology in Manufacturing

RFID in manufacturing completes the operation of automated production and identification, tracks resources, and traces the position of products in production lines to decrease labor costs and error rates and to maximize efficiency. Information in RFID tags can be read first from the RFID reader, and product processing companies can create necessary information from relevant jurisdictions, such as technology for integrated processing, unit date of processing, processing stage, additives used, packing processing, accurate selection by weight on electronic tags, etc.

Sources of information are stored on electronic tags after the processing company's data is faithful. The end user of the retailer or wholesale vendor can then easily recognize the relevant information by looking up the product information through the query terminal. The quality and transparency of the product make it easy to track accidents. Table 1.1 shows the benefits and problems of RFID across applications.

References

1. Ozguven, E. E., & Ozbay, K. (2015). An RFID-based inventory management framework for emergency relief operations. *Transportation Research Part C: Emerging Technologies, 57*, 166–187.
2. Cai, H. B., Andoh, A. R., Su, X., & Li, S. (2014). A boundary condition based algorithm for locating construction site objects using RFID and GPS. *Advanced Engineering Informatics, 28*(4), 455–468.
3. Zhong, R. Y., Dai, Q. Y., Qu, T., Hu, G. J., & Huang, G. R. (2013). RFID-enabled real-time manufacturing execution system for mass customization production. *Robotics and Computer-Integrated Manufacturing, 29*(2), 283–292.
4. Zhong, R. Y., Li, Z., Pang, L. Y., Pan, Y., Qu, T., & Huang, G. Q. (2013). RFID-enabled real-time advanced planning and scheduling shell for production decision making. *International Journal of Computer Integrated Manufacturing, 26*(7), 649–662.
5. Wong, W. K., Guo, Z. X., & Leung, S. Y. S. (2014). Intelligent multi-objective decision-making model with RFID technology for production planning. *International Journal of Production Economics, 147*, 647–658.
6. Guo, Z. X., Ngai, E. W. T., Yang, C., & Liang, X. D. (2015). An RFID-based intelligent decision support system architecture for production monitoring and scheduling in a distributed manufacturing environment. *International Journal of Production Economics, 159*, 16–28.
7. Qu, T., Yang, H. D., Huang, G. Q., Zhang, Y. F., Luo, H., & Qin, W. (2012). A case of implementing RFID-based real-time shop-floor material management for household electrical appliance manufacturers. *Journal of Intelligent Manufacturing, 23*(6), 2343–2356.

8. Zhang, F. Q., Jiang, P. Y., Zheng, M., & Cao, W. (2013). A performance evaluation method for radio frequency identification-based tracking network of job-shop-type work-in-process material flows. *Proceedings of the Institution of Mechanical Engineers, Part B: Journal of Engineering Manufacture, 227*(10), 1541–1557.
9. Wu, C. H., Ho, G. T. S., Yung, K. L., Tam, W. W. Y., & Ip, W. H. (2018). An RFID-based fallen object detection system: A case study of Hong Kong's light rail system. *IEEE Journal of Radio Frequency Identification, 2*(2), 55–67.
10. Tang, T., Lin, T., Jie, W., & Ming, C. (2018). Full realization of digitalization is the only way to intelligent manufacturing. *Chinese Journal of Mechanical Engineering, 29*, 366–377.
11. Alfian, G., Rhee, J., Ahn, H., Lee, J., Farooq, U., Ijaz, M. F., & Syaekhoni, M. A. (2017). Integration of RFID, wireless sensor networks, and data mining in an e-pedigree food traceability system. *Journal of Food Engineering, 212*, 65–75.
12. Wang, S., Wan, J., Zhang, D., Li, D., & Zhang, C. (2016). Towards smart factory for Industry 4.0: A self-organized multi-agent system with big data based feedback and coordination. *Computer Networks, 101*, 158–168.
13. Blanchet, M., Rinn, T., von Thaden, G., & de Thieulloy, G. (2014). Industry 4.0: The new industrial revolution. How Europe will succeed. In *Think act*. Munich: Roland Berger Strategy Consultants GmbH.
14. Ma, H., Wang, Y., & Wang, K. (2018). Automatic detection of false positive RFID readings using machine learning algorithms. *Expert Systems with Applications, 91*, 442–451.
15. Wang, C., Jiang, P., & Ding, K. (2017). A hybrid-data-on-tag-enabled decentralized control system for flexible smart workpiece manufacturing shop floors. *Proceedings of the Institution of Mechanical Engineers, Part C: Journal of Mechanical Engineering Science, 231*(4), 764–782.
16. Hashem, I. A. T., Yaqoob, I., Anuar, N. B., Mokhtar, S., Gani, A., & Khan, S. U. (2015). The rise of big data on cloud computing: Review and open research issues. *Information Systems, 47*, 98–115.
17. Ertek, G., Chi, X., & Zhang, A. N. (2017). A framework for mining RFID data from schedule-based systems. *IEEE Transactions on Systems, Man, and Cybernetics: Systems, 47*(11), 2967–2984.
18. Feng, J., Li, F., Xu, C., & Zhong, R. Y. (2018). Data-driven analysis for RFID-enabled smart factory: A case study. *IEEE Transactions on Systems, Man, and Cybernetics: Systems*, 1–8. https://doi.org/10.1109/TSMC.2018.2882838.
19. Bu, F. (2018). A data mining framework for massive RFID data based on Apriori algorithm. *Journal of Physics: Conference Series, 1087*, 022020.
20. Hao, H., & Zhao, B. (2015). Application of RFID technology in development of warehouse logistics management system by internet of things. *International conference on education, management, commerce and society (EMCS-15). Atlantis Press*. https://doi.org/10.2991/emcs-15.2015.89.
21. Ertek, G., Chi, X., & Zhang, A. (2017). A framework for mining RFID data from schedule-based systems. *IEEE Transactions on Systems, Man, and Cybernetics: Systems, 47*(11), 2967–2984
22. Liu, Y., et al. (2012). Mining frequent trajectory patterns for activity monitoring using radio frequency tag arrays. *IEEE Transactions on Parallel and Distributed Systems, 23*, 2138–2149.
23. Shen, B., Zheng, Q., Li, X., & Xu, L. (2015). A framework for mining actionable navigation patterns from in-store RFID datasets via indoor mapping. *Sensors (Basel), 15*(3), 5344–5375.
24. Kang, H. S., Lee, J. Y., Choi, S., et al. (2016). Smart manufacturing: past research, present findings, and future directions. *International Journal of Precision Engineering and Manufacturing-Green Technology, 3*, 111–128.
25. Xia, F., Yang, L. T., Wang, L., & Vinel, A. (2012). Internet of things. *International Journal of Communication Systems, 25*(9), 1101–1102.

26. Atzori, L., Iera, A., & Morabito, G. (2017). Understanding the internet of things: Definition, potentials, and societal role of a fast evolving paradigm. *Ad Hoc Networks, 56*, 122–140.
27. Atzori, L., Iera, A., & Morabito, G. (2010). The Internet of things: A survey. *Computer Networks, 54*(15), 2787–2805.
28. Jia, X., Feng, O., Fan, T., & Lei, Q. (2012). RFID technology and its applications in internet of things (IoT). In *Proceedings of the 2nd IEEE International Conference on Consumer Electronics, Communications and Networks (CECNet)*, Yichang, China, Apr. 21–23, 2012, pp. 1282–1285.
29. Liu, C. H., Yang, B., & Liu, T. (2014). Efficient naming, addressing and profile services in internet-of-things sensory environments. *Ad Hoc Networks, 18*, 85–101.
30. Yaqoob, I., Ahmed, E., Hashem, I. A. T., Ahmed, A. I. A., Gani, A., Imran, M., & Guizani, M. (2017). Internet of things architecture: Recent advances, taxonomy, requirements, and open challenges. *IEEE Wireless Communications, 24*(3), 10–16.
31. Da Xu, L., He, W., & Li, S. (2014). Internet of things in industries: A survey. *IEEE Transactions on Industrial Informatics, 10*(4), 2233–2243.
32. Mashal, I., Alsaryrah, O., Chung, T. Y., Yang, C. Z., Kuo, W. H., & Agrawal, D. P. (2015). Choices for interaction with things on internet and underlying issues. *Ad Hoc Networks, 28*, 68–90.
33. Kumar, N. M., & Mallick, P. K. (2018). The internet of things: Insights into the building blocks, component interactions, and architecture layers. *Procedia Computer Science, 132*, 109–117.
34. IIC Working TWG and SWG. (2015). *Industrial internet reference architecture*. Needham: Industrial Internet Consortium.
35. Adolphs, P., Bedenbender, H., Ehlich, M., Epple, U., Hankel, M., Hoffmeister, M., et al. (2015, July). GMA status report: Reference Architecture Model Industrie 4.0 (RAMI 4.0). *VDI/VDE-Gesellschaft Mess- und Automatisierungstechnik*. Retrieved from https://www.zvei.org/fileadmin/user_upload/Presse_und_Medien/Publikationen/2016/januar/GMA_Status_Report__Reference_Archtitecture_Model_Industrie_4.0__RAMI_4.0_/GMA-Status-Report-RAMI-40-July-2015.pdf
36. Oprea, A., Barsan, N., Weimar, U., Bauersfeld, M. L., Ebling, D., & Wöllenstein, J. (2008). Capacitive humidity sensors on flexible RFID labels. *Sensors and Actuators B: Chemical, 132*(2), 404–410.
37. Delen, D., Sharda, R., & Hardgrave, B. C. (2011). The promise of RFID-based sensors in the perishables supply chain. *IEEE Wireless Communications, 18*(2), 82–88.
38. Barreto, L., Amaral, A., & Pereira, T. (2017). Industry 4.0 implications in logistics: An overview. *Procedia Manufacturing, 13*, 1245–1252.
39. Zhong, R. Y., et al. (2017). Intelligent manufacturing in the context of industry 4.0: A review. *Engineering, 3*(5), 616–630.
40. Tao, F., Cheng, Y., Xu, L. D., Zhang, L., & Li, B. H. (2014). CCIoT-CMfg: Cloud computing and internet of things based cloud manufacturing service system. *IEEE Transactions on Industrial Informatics, 10*(2), 1435–1442.
41. Bi, Z., Xu, L. D., & Wang, C. (2014). Internet of things for enterprise systems of modern manufacturing. *IEEE Transactions on Industrial Informatics, 10*(2), 1537–1546.
42. Lu, B. H., Bateman, R. J., & Cheng, K. (2006). RFID enabled manufacturing: Fundamentals, methodology and applications. *International Journal of Agile Systems and Management, 1*(1), 73–92.
43. Wang, Y. M., Wang, Y. S., & Yang, Y. F. (2010). Understanding the determinants of RFID adoption in the manufacturing industry. *Technological Forecasting and Social Change, 77*(5), 803–815.
44. Huang, G. Q., Zhang, Y. F., Chen, X., & Newman, S. T. (2008). RFID-enabled real-time wireless manufacturing for adaptive assembly planning and control. *Journal of Intelligent Manufacturing, 19*(6), 701–713.
45. Liu, W. N., Zheng, L. J., Sun, D. H., Liao, X. Y., Zhao, M., Su, J. M., et al. (2012). RFID-enabled real-time production management system for Loncin motorcycle assembly line. *International Journal of Computer Integrated Manufacturing, 25*(1), 86–99.

46. Zhang, Y., Zhang, G., Wang, J., Sun, S., Si, S., et al. (2014). Real-time information capturing and integration framework of the internet of manufacturing things. *International Journal of Computer Integrated Manufacturing, 28*(8), 1–12. https://doi.org/10.1080/0951192X.2014.900874.
47. Dai, Q. Y., Zhong, R. Y., Huang, G. Q., Qu, T., Zhang, T., & Luo, T. Y. (2012). Radio frequency identification-enabled real-time manufacturing execution system: A case study in an automotive part manufacturer. *International Journal of Computer Integrated Manufacturing, 25*(1), 51–65.
48. Zhong, R. Y., et al. (2015). A big data approach for logistics trajectory discovery from RFID-enabled production data. *International Journal of Production Economics, 165*, 260–272.
49. Hofmann, E., & Rüsch, M. (2017). Industry 4.0 and the current status as well as future prospects on logistics. *Computers in Industry, 89*, 23–34.
50. Tajima, M. (2007). Strategic value of RFID in supply chain management. *Journal of Purchasing and Supply Management, 13*(4), 261–273.
51. Landmark, A. D., & Sjøbakk, B. (2017). Tracking customer behaviour in fashion retail using RFID. *International Journal of Retail & Distribution Management, 45*(7/8), 844–858.
52. Madhani, P. M. (2015). Enhancing customer lifetime value in fast fashion retailing with RFID initiatives. *International Journal of Business and Globalisation, 15*(2), 205–237.
53. Zhou, Z., Shangguan, L., Zheng, X., Yang, L., & Liu, Y. (2017). Design and implementation of an RFID-based customer shopping behavior mining system. *IEEE/ACM Transactions on Networking, 25*(4), 2405–2418.
54. Choi, S. H., Yang, Y. X., Yang, B., & Cheung, H. H. (2015). Item-level RFID for enhancement of customer shopping experience in apparel retail. *Computers in Industry, 71*, 10–23.
55. Niggemann, O., Biswas, G., Kinnebrew, J. S., Khorasgani, H., Volgmann, S., & Bunte, A. (2015). Data-driven monitoring of cyber-physical systems leveraging on big data and the internet-of-things for ciagnosis and control. In *CEUR Workshop Proceedings*, Vol. 1507, pp. 185–192.
56. Wang, S., Wan, J., Li, D., & Zhang, C. (2016). Implementing smart factory of Industrie 4.0: An outlook. *International Journal of Distributed Sensor Networks, 12*(1), 3159805.
57. Rüßmann, M., Lorenz, M., Gerbert, P., Waldner, M., Justus, J., Engel, P., & Harnisch, M. (2015). *Industry 4.0: The future of productivity and growth in manufacturing industries. Boston Consulting Group, 9*(1), 54–89.
58. Temidayo, O., Oke, A. E., & Aigbavboa, C. (2018). Fusing RFID with mobile technology for enhanced safety of construction project team members. *Proceedings of the International Conference on Industrial Engineering and Operations Management, 2018*, 312–321.
59. Jiang, P., Sun, P., Kai, D., & Wang, C. (2015). Radio frequency identification-based graphical formalized deduction method for process tracking and its industrial applications. *Chinese Journal of Mechanical Engineering, 51*(20), 9–17.
60. Alyahya, S., Wang, Q., & Bennett, N. (2016). Application and integration of an RFID-enabled warehousing management system – A feasibility study. *Journal of Industrial Information Integration [Internet], 4*, 15–25.
61. Wang, Q., Alyahya, S., Bennett, N., & Dhakal, H. (2015). An RFID-enabled automated warehousing system. *International Journal of Materials, Mechanics and Manufacturing, 3*(4), 287–293.

Chapter 2
Data Fusion-Based AI Algorithms in the Context of IIoTS

Raluca Maria Aileni, Suciu George, Sever Pasca, and Valderrama Sukuyama Carlos Alberto

Industrial Internet of Things: Overview in the Context of Industry 4.0 and Massive Data Exploitation

In general, the IoT [1] concept enables Internet access to any type of electronic device (thing). Thus, each "thing" will have an IP (Internet Protocol) address and will be addressable using the standard Internet technology like DNS (domain name system). In an automation context, the concept of the Internet of Things will be a device or functionality that we can find in standard ISA-95 automation implementations.

Automation applications of the Internet of Things typically include sensors, actuators, programmable logic controllers (PLCs), control loops, data analytics, control models, simulators, optimizers, etc. Thereby, the concept can be viewed as a physical device or as a functionality that is computed in a software system executed on any type of device having sufficient computational resources. Currently, there is no consistent technology that can be attributed to a device belonging to the Internet of Things. Several standardization efforts are underway. For example, ITU (the International Telecommunication Union) has a standardization working group SG20 on "IoT and its applications including smart cities and communities" [2], and IEEE (the Institute of Electrical and Electronics Engineers) has an ongoing IoT Ecosystem study and the IEEE P2413 draft standard for an architectural framework for the Internet of Things [2].

Nowadays, IoT (Internet of Things) has developed into a core technology of hyper-connected society that connects and interacts with all things (elements/objects), spaces, and environments in the real world. In accordance with this trend,

R. M. Aileni (✉) · S. George · S. Pasca
Politehnica University of Bucharest, Bucharest, Romania

V. S. Carlos Alberto
University of Mons, Mons, Belgium

© Springer Nature Switzerland AG 2020

G. R. Kanagachidambaresan et al. (eds.), *Internet of Things for Industry 4.0*, EAI/Springer Innovations in Communication and Computing,
https://doi.org/10.1007/978-3-030-32530-5_2

most of the industrial fields such as electricity, transport, medical, retail, and manufacturing are trying to create advanced services and new businesses through fusion with IoT technology. Especially in the manufacturing industries, there are an increased number of smart factories that realize that intelligent production automation can be achieved by using IoT technology.

The Industrial Internet of Things [3] gives us a way to obtain proper visibility and insight into the operation of the company and assets by integrating software, backend cloud computing, and storage systems. Shortly, we can say that the Industrial Internet of Things offers a method of business operational processes transformation using as feedback the results given by big data sets interrogation through advanced analytics.

The Internet of Things (IoT) in Industry 4.0 [4] can support collaboration and connection between objects automatically. Therefore, with the increasing number of involved devices, IoT systems may consume large amounts of energy, and, taking it into consideration, relevant energy efficiency issues have recently been attracting much attention from both academia and industry.

Recently, IoT has evolved [5] into edge computing technology that enhances data processing and analysis by decentralizing the computing power that was performed in the central cloud to an end device. Since edge computing is a structure suitable for a manufacturing environment requiring real-time processing of data and fast response, many studies have considered edge computing as an optimized platform technology for constructing a smart factory.

Industry 4.0 has as a reference the 4th Industrial Revolution, and the introduction of Internet technology has as a reference the manufacturing industry, having the purpose to obtain more intelligent factories, increased adaptability, and the efficiency of the resources.

The future of Industry 4.0 is to go beyond new computer integrated manufacturing techniques. Here, we can talk about nine technologies that are transforming industrial production: simulation, autonomous robots, augmented reality, the Industrial Internet of Things (IIoT), Digital Twin, the cloud, cybersecurity, additive manufacturing, horizontal/vertical system integration, and, in the end, big data analytics. The industrial IoT links all of the components, equipment, and products with embedded computing and sensors of the company. One of the essential components of Industry 4.0 is the fusion of the physical and virtual world. This fusion can only be made possible by cyber-physical systems (CPSs).

To continue with Industry 4.0, the most important feature is represented by smart factories. A smart factory is, in fact, a factory where CPSs and IoT help people and machines in the achievement of their tasks.

Some sub-processes [6] for a smart factory:

M2M (machine-to-machine) communication via IoT
Consistent communication between the sensor and the cloud
Robotics and innovative drive technologies integration

Radio-frequency identification (RFID) as the basis for parts tracking and intelligent products

The smart factory aims to [7] build an intelligent collaborative system in which a variety of detecting devices (sensors) and machines that have been physically connected to the manufacturing site is objectified and interconnected via the telecommunication network to share the production process freely and control it autonomously.

Thus, to accomplish this [8], it is important to help the platform structure and communication environment in which data, produced from production devices or IoT sensors, can be vertically and horizontally exchanged, integrated, and analyzed with other devices and upper systems.

The rest of the paper is organized as follows: In Sect. 2 the IIoT systems are presented, which are based on a modular architecture in Industry 4.0, followed by the section portraying the cybersecurity layers of software and hardware. Section 4 depicts notions about big data fusion in the context of IoT systems based on smart sensors, pursued by the fifth section introducing concepts about IIoTS communication protocols and cloud computing services. Thus, Sect. 6 enounces ideas about data fusion based on advanced AI algorithms, and henceforth, the last section illustrates the conclusion of this paper and envisions the future work.

IIoT Systems Based on Modular Architecture for Industry 4.0

Industry 4.0 is related to the IoT (Internet of Things) (Fig. 2.1), CPSs (cyber-physical systems), ICT (information and communications technology), EA (enterprise architecture), and EI (enterprise integration). The main objective of Industry 4.0 is to accomplish a large amount of operational effectiveness and productiveness and also a more elevated level of automatization [9]. The attributes of the Industry 4.0 [10, 11] are digitization, optimization, customization of production, automation and adaption, HMI (human-machine interaction), value-added services and businesses, and automatic data exchange and communication. Such attributes are exceptionally related to technologies and modern algorithms and show that the Industry 4.0 represents an industrial activity which has the purpose to add significant value and information management.

IIoT (Industrial Internet of Things) semantically depicts an applied science movement, whereas the Industry 4.0 is highly connected with the anticipated financial impact. IIoT leads to the Industry 4.0 [12]. IIoT is not exclusively the network of the physical objects in the industry; yet, it additionally incorporates the computerized representations of the products, procedures, and factories, such as the 3D model or physical behavior models of the machines. In the final report of Industry 4.0 Working Group, published in 2013 [13], industry experts and academicians have been attempting to completely understand the Industry 4.0

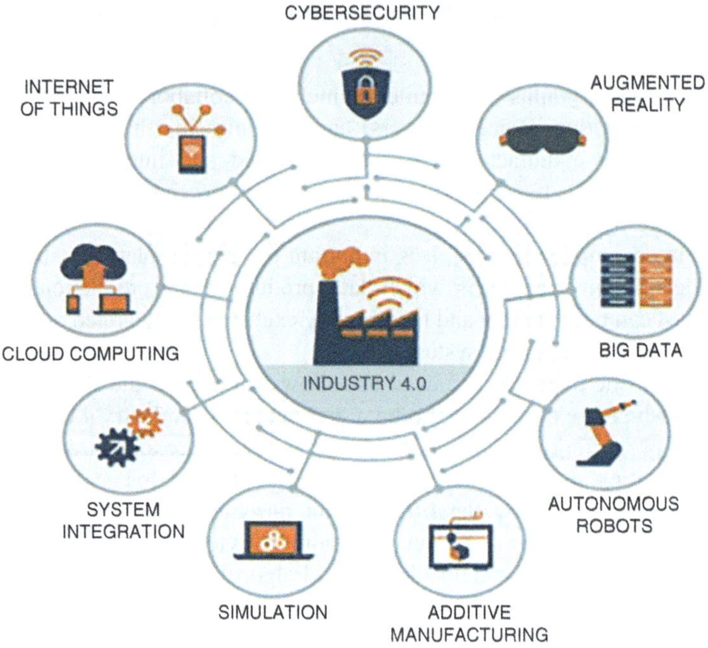

Fig. 2.1 Industry 4.0 [14]

repercussions for the manufacturing process. The most important repercussions in this context are for manufacturing IT systems.

The word "Industry 4.0" relates to [15] production's growing digitalization. Experts look forward to a change from centrally managed manufacturing systems with fixed process flows to cyber-physical system (CPS) manufacturing systems for the future.

The possibilities to customize [16] the production process are quite limited due to the static structure of factories. Based on this shortcoming, the idea of a modular, decentralized, and digital system arose. The modularity shall provide the flexibility to design and implement customized system structures. Another benefit would be the possibility to extend the system with other modules directly.

Advanced digital technology [17] is used in the production process. Industry 4.0 is going to transform the entire production domain and it will lead to increased efficiency. The traditional production is going to change the relationships among producers, suppliers, and customers. The connection between humans and machines is going to be changed as well. At this point, there are nine technology trends that design the building blocks of Industry 4.0:

- *Big Data and Analytics* – In the context of Industry 4.0, the collection and complete evaluation of data from numerous distinct sources – the equipment used in production and the enterprise- and customer management-related systems – are going to become standard in order to support taking real-time decision.

- *Autonomous Robots* – Robots will finally interact with one another. They are going to work safely with humans, learning from them simultaneously. These robots won't be expensive, and they will have a greater potential compared to the robots currently used in the manufacturing industry.
- *Simulations* – They are going to be utilized furthermore in plant operations in order to take advantage of real-time data, mirror the physical world, and create a virtual model which can incorporate machines, products, and humans. This will allow workers to test and develop the machine settings used in the next product in line in the virtual created world before the physical modification, thereby decreasing the setup time of the machines and increasing their quality.
- *Horizontal and Vertical System Integration* – Using the Industry 4.0, companies, their departments, functions, and different capabilities will be more cohesive, as cross-company, common data-integration networks emerge and facilitate entirely automated value chains.
- *Cybersecurity* – Alongside the increased connectivity and usage of standard communication protocols that work with Industry 4.0, the demand of protecting crucial industrial systems and manufacturing lines from cybersecurity hazards is completely growing. As a result, there are some essential aspects, such as secure and decent communications, complex identity, and access management of users and machines.
- *The Cloud* – More production-related undertakings are going to necessitate increased data sharing across sites and company bounds. Simultaneously, the efficiency of cloud technologies will advance, achieving reaction times of only several milliseconds. As a result, machine data and performance will increasingly be deployed to the cloud, enabling more data-driven services for production systems.
- *Additive Manufacturing* – Companies have just begun to adopt additive manufacturing. A good example of this concept would be 3D printing, which is used to produce individual components and to prototype. With Industry 4.0, these additive-manufacturing methods can be widely adopted to produce small batches of customized products that offer construction advantages.

Three frameworks can further specify key elements of Industry 4.0: the smart product, the smart machine, and the augmented operator.

The smart product's guiding concept is to extend the workpiece's function to an active aspect of the scheme. The products obtain a memory on which operational information and specifications are stored as an employee construction plan directly.

Smart machine defines the cyber-physical production systems (CPPS) process of machines. Decentralized self-organization enabled by CPS will replace the traditional manufacturing hierarchy. They portray autonomous parts with local control intelligence that can interact through open networks and semantic descriptions to other field systems, manufacturing modules, and products. This enables machines to organize themselves within the manufacturing network.

Production lines [18] will become so flexible and modular, that under situations of highly flexible mass manufacturing, it can be designed even in the smallest lot size.

The augmented operator, in the challenging setting of fully modular manufacturing systems, aims at the worker's technological support. Industry 4.0 does not tend to migrate to evolving manufacturing without the worker.

In the Industry 4.0 [19], IoT is required to offer auspicious solutions for the operation and role of the industrial systems. Industry 4.0 is an initiative with technological innovations such as cloud computing, electric vehicles (EV), the Internet of things, artificial intelligence, and cyber-physical systems. Industry 4.0 has attracted attention from governments, researchers, and industries.

Cybersecurity Aspects Correlated with the Software and Hardware Layers

An IIoT system [20] is designed to exist for a long time. In an IIoT infrastructure, a significant number of devices are monitoring and controlling a process control entity. More IoT-based industrial entities or plants may interact with each other involving these IoT smart devices. In real life, IIoT can be applied in smart cities for transportation, smart homes, wearables, emergency services, air and energy monitoring systems, and oil and gas industry. As the number of IIoT devices increases, the number of open connections increases as well and so the vulnerability to attacks. As a known example of weakness, devices placed in the exterior are exposed to tampering. Regarding devices which monitor a specific area, someone can control them to track something else.

The NIST (the National Institute of Standards and Technology) created the Cyber Security Framework (NIST, 2014), which organizes the following categories of cybersecurity activities:

- *Identify* – the growth of the organizational knowledge on how to control cybersecurity risks of the systems, data, and assets.
- *Protect* – the development and the implementation of the appropriate protection in order to ensure delivery of critical services.
- *Detect* – the development and the implementation of the appropriate activities in order to identify the circumstance of a cybersecurity event.
- *Respond* – the development and the implementation of the appropriate activities in order to act when a cybersecurity event is detected.
- *Recover* – it enables separation of this framework from the others.

These categories [21] can provide features for an organization to handle risk decisions and address threats and incident management. The Cyber Security Framework considers the importance of recovery planning and suggests the following for recovery of any attacks caused by cyberattacks. The IIoTS has building support for the sensor-cloud. The sensors are integrated into this sensor-cloud, being organized as wireless sensor networks (WSNs). For better wireless network security,

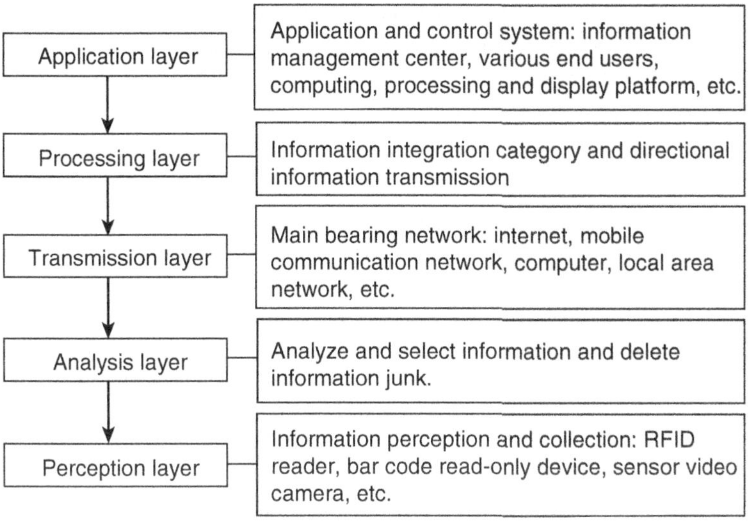

Fig. 2.2 Software layers [23]

the solution is a network node trust with several detection routes. In an IIoT environment, the sensor-cloud assures communication of the WSN. The information gathered from the sensors is sent to the cloud. Next, data is stored and processed in more data centers, which are mainly created for this purpose. For users, this processed data can be sent on demand from the cloud.

Too many devices included in an IIoT infrastructure [22] can slow the update of software defenses, as well as hardware security measures. Known defensive solutions in the IIoT domain are replacing the system components or shutting down the compromised system. This is not enough for IIoTS security.

An IIoT system uses wireless communication, having a sensor/device layer; a communication layer, i.e., the medium where wireless signals propagate; and the user application layer (Fig. 2.2), where data is received. Changing the position of an IIoT surveillance camera or taking (a camera) and placing it somewhere else. Stealing components of a device results in data disclosure. As a solution, an IIoT device can be equipped with motion or temperature sensors, e.g., gyroscopes.

Unauthorized access has the main purpose of finding vulnerabilities, including passive activities like probe computers or network, capturing, analyzing, and disclosure of packet content.

Attacks on the Perception Layer

Several sensors such as WSN, Zigbee, RFID, infrared, QR code, and other kinds are used to collect information (temperature, humidity, force, pH levels, pressure, vibration, etc.) which is then transmitted through the network/communication layer to the central information processing unit.

This layer is the most susceptible to hardware attacks. The attacker is required to be in the network or physically close to the nodes of the system. Here are some common attacks on this layer:

- *Fake node injection* – an infected node is placed between the nodes of the network, making an opening for the attacker into the network and hence gaining the ability to control all the data flow. This intervention could cause the node to stop transmitting data and, in consequence, destroy the whole network.
- *Hardware tempering* – the attacker changes parts of the node's hardware or replaces the whole node and therefore gains information regarding the network, like communication key, cryptographic key, routing table, etc. This puts into danger even the higher layers of the network.
- *Malicious code injection* – the transmission is halted because of DoS attacks in WSN or virus on the nodes. The network loses resources and makes the services unavailable.
- *Sleep denial attack* – nodes located in more remote places in the network generally run on batteries and enter a sleep mode when they are not in use in order to increase their battery life. This type of attack implies that wrong input is fed to the nodes to prevent them from entering the sleep mode, which leads to power being consumed until the nodes shut down.
- *WSN node jamming* – wireless sensor networks perform on radio frequency. In order to deny service and communication between nodes, the attacker sends noise signals over the network or jams the signals of WSN.
- *RF interference of RFIDs* – RFIDs work on radio frequency as well. The attacker interferes with the RFID signal by sending noise over the network and creating a distortion in the node communication.

Attacks on the Communication Layer

The information is transmitted successfully in this layer through various mediums like RFTD, satellite, infrared, and Wi-Fi units, with regard to the type of sensors and the sensitivity of data.

From a [24, 25] network point of view, these are some examples of known active attacks in IIoTS:

- Packet dropping.
- RFID spoofing – intrusion in the network and manipulation of data.
- RFID unauthorized access – the data in nodes can be easily modified (read, write, delete), as there is no secured authentication system.
- Fraudulent packet injection attacks – i.e., replay captured packet, sending false packets that cause malfunctioning of specific tasks.
- Sinkhole attacks – they guide all signals from WSN nodes to a single common point. Such attacks nullify the safety of data and drop all the packets rather than deliver them to the destination.

- Routing information attacks – instantaneous attacks where the intruder can coil up the system and make routing loops, send false alerts, shorten or amplify source courses, and permit or drop movement, all by spoofing, replaying, or changing routing data.
- Man-in-the-middle attacks – the intruder meddles with two sensor nodes in order to gain access to confined information. This kind of assaults violates the privacy of the nodes and doesn't require the intruder to be physically present on the site; using the communication protocol of the IoT is enough.

Attacks on the Processing Layer

This layer blends the communication layer and the application layer. Here is where all the cloud and intelligent computing are done. The performance of this layer includes storing the data from lower-level layers into the database and managing the service. It can as well compute information and process data automatically.

The most common type [25] of attacks that occur in this layer are cloud attacks because of the fact that, during this phase, the data is sent to the cloud. These are some frequent attacks that affect the network:

- *Shared resources* – the attacker is able to monitor all the resources being shared between virtual machines using convert channels; thus the information risks to be compromised.
- *Third-party relationships* – third-party web services combine more than one source, leading to an increase in the number of security issues like network and data security.
- *Underlying infrastructure security* – lower-level layers can't be accessed by the developer, so the underlying layer's security becomes the responsibility of the provider. Though the developer could implement a secure application, its security is still vulnerable due to lower layers.

Attacks on the Application Layer

This layer grants services as per usual demand. The processed information coming from lower layers is involved in producing services that are useful for end users. It provides a platform for such applications from which users could benefit in many ways:

- *Malware attacks (viruses, worms, Trojan Horse, spyware)* – they destroy system files, tamper with data, and control the computer. Either it is disguised in software or the attacker injects it.

- *Denial-of-service (DoS) attacks* – they cause depletion on the system's energy, slow processing, until the computer blocks or shuts down. Distributed DoS (DDoS) is designed to attack more devices which will send a high volume of traffic to networks that will consume computer resources.
- *Phishing attacks* – leakage of data.
- *Malicious scripts* – they cause the system to close up and, by running executable active-x scripts, the data is stolen.

There are measures deployed in IIoTS as following:

- Embedded security: device authentication and authorization to access the network for certain operations using authentic software, specially designed for that network.
- Data integrity: applying cryptographic hash functions on the data so it isn't tempered with when it reaches the receiving end.
- Fragmentation redundancy scattering (FRS): the sensitive data on the cloud is separated into fragments and stored on distinct servers. The fragments don't contain any meaningful information by themselves, fact which minimizes the risk of leakage.
- In IIoTS, confidentiality and integrity could be solved with data encryption.
- An IIoTS firewall includes rules and attacks signatures for filtering the traffic; they are always updated to repair new issues regarding vulnerabilities or attacks (i.e., patches). Unexpected behavior in IIoTS.
- Anomalies in datasets can [25] be classified as such:
  - Contextual anomaly – a command with its data value predefined range is executed in a no-show time duration.
  - Collective anomaly – multiple TCP (Transmission Control Protocol) requests.
  - In networks: attack patterns or signatures – a new attack can be found if it corresponds with the pattern; this method is integrated into firewalls; analysis of behaviors that differ from previous ones sometimes brings false positives.

IIoT comes [23] with the sensor-cloud paradigm and the trust-based mechanisms: independent sensor-cloud (ISC), collaborative sensor-cloud (CSC), and mutual sensor-cloud (MSC).

In ISC, a sensor network assigns to each node a trust value and a trust value threshold; a cloud assigns to the data center a trust value and a trust value threshold.

In CSC, a sensor node is assigned to a trust value only by its network, while a trust value threshold is assigned by both the sensor network and the cloud.

In MSC, sensor nodes' trust values and trust value thresholds are assigned by their sensor network. Data centers' trust values and trust value thresholds are assigned by the cloud. However, here, apart from the other mechanisms, trust value thresholds for data centers are mutual between WSN, cloud, and users.

Big Data Fusion in the Context of IIoT Systems Based on Smart Sensors

In the last few years, sensors [26] have been utilized all over the place, and nowadays, they have the potential to impersonate human beings very tightly. Data fusion enables this to happen, which influences a microcontroller to fuse the individual information gathered from different sensors in order to obtain a more precise and valid perspective on the information than the one collected from each discrete sensor all alone.

A basic example of sensor fusion is represented by an e-compass, in which the mix of a 3D magnetometer and a 3D accelerometer gives functionality to the compass. Furthermore, more elaborated fusion technologies give clients an innovative and more improved experience, by consolidating 3D accelerometers, 3D gyroscopes, and 3D magnetometers (which estimate the constituent elements of the magnetic field in a specific direction, in comparison with the spatial orientation of a given device). Every one of these sensor types provides proper functionality and has the following constraints:

- Accelerometers: X, Y, and Z axis linear movement detection, however sensible to vibrations
- Gyroscope: presents a rotational sensor (pitch, roll, and yaw), but zero bias drift
- Magnetometer: X, Y, Z axis magnetic field detection, but sensitive to magnetic interference

Another example of data fusion is a gaming platform that can recognize feelings electronically by keeping under observation and gathering data from physiological factors and states:

- Muscle relaxation (MR), through a pressure sensor
- Heart rate variability (HRV), through a two-electrode ECG on a chip
- Sweat (S), through capacitive sensor
- Attitude (A), through an accelerometer that monitors the state of relaxation of a human being
- Muscle contraction (MC), through a pressure sensor

Information [27] collected from the sensors can be utilized for various services, giving several advantages and simplifying people's lives. Another utilization is for data mining and other use cases that increase security and privacy problems regarding IoT technologies. More capabilities can be assembled with data fusion and REC technologies. IoT will cover each part of our lives in under 10 years, sensor fusion having the main impact of this spreading phenomenon. Big data processing has a significant impact nowadays, with mechanisms and tools that produce various services.

IoT creates a large amount of information which would be less valuable except the case in which people can learn by utilizing them. Data fusion, one of the critical challenges in IoT, creates innovative services. In IoT environments, data fusion represents a big challenge that should be tackled to create innovative services. In the particular case of smart cities communities, when 50–100 billion gadgets start sensing [28], a fusion will be fundamental. Data fusion can be applied at two levels: cloud and network levels.

Data fusion can be described as a processing method that connects, compounds, reunites, and incorporates information from various sources. Moreover, it contributes to developing a context-awareness model that permits comprehension of the situational context. The sensor information filter emphasizes the necessity of avoiding vast volumes of information to be transmitted over the network.

Recent buyers of the IIoT [29] have recognized competitive upper hands and innovative business models for expanding the income, reduce prices, and also enhance the customer service and support. Expressions such as artificial intelligence, predictive maintenance, and augmented and virtual reality are not anymore just favorite expressions. The ideas, innovations, and concepts provided by the adopters are being embraced and connected to these industries each day. Nevertheless, IIoT adoption challenges continue to exist.

IIoT (Industrial Internet of Things) represents a system which associates and incorporates OT (operational technology) environments, including ICS (industrial control systems) with business frameworks, processes, and analytics. IIoT systems, besides ICS and OT, are associated widely to different frameworks and individuals. Also, they utilize sensors and actuators that interface with the physical world; uncontrolled changes can produce precocious circumstances. IIoT has the following advantages: a large number of sensors can gather and process the information; based on what the information reveals, the system will react. The precise connectivity increases the risk of cyberattacks by the individuals who might need to cut down the framework.

IIoT systems present [30] a framework of the Industry 4.0, in which research topics are categorized into the smart design, smart machines, smart monitoring, smart control, and smart scheduling. Smart design is evolving with the rapid development of new technologies such as virtual reality (VR) and augmented reality (AR), so the traditional design will be upgraded. Design software such as computer-aided design (CAD) and computer-aided manufacturing (CAM) can interact with physical smart prototype systems in real time, enabled by three-dimensional (3D) printing integrated with CPSs and AR. In Industry 4.0, intelligent machines are made using smart robots and other types of smart objects capable of perceiving and interacting in real time. Smart monitoring is an essential aspect for the operations, maintenance, and optimal scheduling of Industry 4.0 manufacturing systems. The widespread deployment of various types of sensors makes it possible to achieve smart monitoring. For example, data and manufacturing factors such as temperature, electricity consumption, vibrations, and speed can be obtained in real time. In Industry 4.0, high-resolution and adaptive production control (i.e., smart control) can be achieved by developing cyber-physical production-control systems. Smart control is mainly executed to physically manage various smart machines or tools through a cloud-enabled platform. End users can switch off a machine or a robot via their smartphones. The smart scheduling layer mainly includes advanced models and algorithms to draw on the data captured by sensors. Data-driven techniques and advanced decision architecture can be used for intelligent scheduling. For example, to achieve real-time, reliable scheduling and execution, distributed smart models using a hierarchical interactive architecture can be used.

The replacement of human [30] resources with intelligent systems requires the adaptation of these systems so that they can safely monitor the industrial process, which implies the development of specialized IIoT systems. Although there are many techniques to identify the issues that can be selected, intelligent production puts a strong emphasis on predictive capacity development across the factory. The ability to accurately predict problems can ensure the availability of equipment and availability for demand-side supply offers while optimizing energy consumption and the cost of life equipment by supervising sensor systems and depending on the sensor parameters to make individual decisions.

The topology of an industrial network contains standard components, such as devices, systems, and databases, which facilitate the factory data flow. This topology describes a particular type of network that is used in production to orchestrate the production process, known as an automation network. Since the adoption of standards is a significant issue for these types of industrial systems, the topology presented is abstracted at a level that allows hierarchical data flow to be displayed without being burdened with low-level details. Data transmission begins with instruments at the level of low-level recording measurements (e.g., temperature) and culminates in end-user access to relevant information.

The sensors transmit continuous measurements (e.g., room temperature, microphone signal, robot actuator status) to programmable logic controllers (PLCs). PLCs are digital computers logically programmed to automate the production process. The PLC is programmed by automation engineers to evaluate the parameters measured by each sensor and to initiate appropriate actions according to their state. Measurements transmitted to PLCs are persistent in memory at established intervals. This type of data transmission is standard in industrial IT systems and is referred to as serial data, measured or timed data (Fig. 2.3). Subsequently, persistent PLC time data is transferred to an archive at a specified interval (e.g., every 2 hours), the file typically taking the form of a relational database. Once the data persists in the file, computer systems can consume data from the warehouse and generate reports for

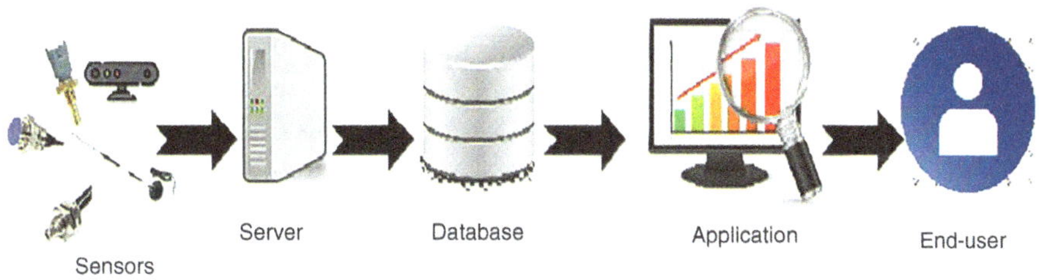

Fig. 2.3 General mode of data transmission

end users. Examples of this type of systems are building management systems, air quality monitoring systems, and industrial production systems. Access to data in the archive can be done using standard database and I/O interfaces. However, big data may require some processing to be used; for example, the noise of a microphone recorded signal will be eliminated. While accessing archive data allows it to be used in some industrial applications, the high latency characteristics of archiving in automation networks means that it is not inherently appropriate to real-time industrial applications. If real-time sensor data are to be accessed, there must be a direct communication with the devices (e.g., PLC usage) on the lower levels of an automation network. This can be done using industry protocols and interfaces such as Modbus, LonWorks, BACnet, OLE Process Control (OPC), and MTConnect or where smart sensors are using MQTT, COAP, and HTTP. However, as mentioned above, the adoption of intelligent sensors and emerging technologies in large-scale industrial plants can be restricted by quality regulation and control, as well as the high risks associated with adopting new technologies. Therefore, initial data management and infrastructure requirements for intelligent manufacturing may need to consider how old and emerging technologies can work. An alternative to storing data is also their storage in the cloud.

In order to minimize human intervention in industrial processes, applications should reduce the workload of personnel by making calculations, process the data and display them in a way that the supervisor understands, or implement notification modules considering the data processed to alert authorized personnel or the technical team.

Figure 2.4 illustrates an example of an IIoT system as a real-world network of digital counterparts. Product life cycle management (PLM) provides data about the production line status. Digital factory is a comprehensive model of the real factory that can be used for communication, simulation, and optimization during its life cycle. Software products in the domain of digital factory typically come with different modules enabling functions such as material flow simulation, robot programming, and virtual commissioning. In the context of IIoT, the digital

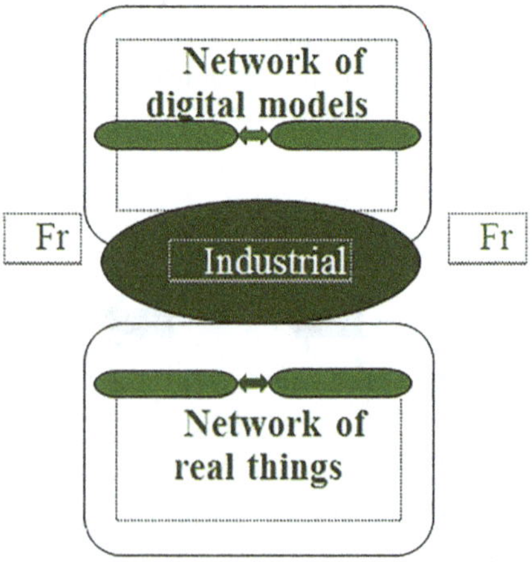

Fig. 2.4 An IIoT system as a real-world network of digital counterparts [28]

factory can be considered as a complement to PLM. While PLM aims to integrate data along the product life cycle, the digital factory incorporates the data of production resources and processes. For the IIoT both are necessary, high-fidelity models of the product and its products based on data from sensors subsequently processed.

IIoTS Communication Protocols and Cloud Computing Services

IIoTS communication [31] was developed for many years. A good example of this could be the vehicular networks where two or more cars can communicate between them with a wireless network connection (WSN). This dynamic communication was named ad hoc networks (VANET), and it was made for customers who wanted a safer car with different functionality such as traffic monitoring system, update, emergency warning, and road assistance. Another part in [32] the automotive domain is cloud computing, based on network solutions. The data collected from external factors was stored in a cloud platform for safety reasons. Part of IIoTS communication [33] is used in the medical domain, more precisely in the healthcare sector, which can help doctors to monitor and control patient's healthcare.

SCADA (Supervisory Control and Data Acquisition) Systems [34] are tasked with monitoring and controlling a network of PLCs (programmable logic controllers) and RTUs (remote terminal units) that measure the performance of local operations, with the use of sensors, and acquire information. It offers real-time updates on the status of its physical processes within its network. A secure communication needs to be established between the devices on the field and the SCADA host in order to transmit the data, to register the updates, and to gain control over the physical processes successfully. This concept is accomplished through an array of specific communication protocols, which carry the information coming from devices on the field all the way to a cloud or local central control center. Vendors started coming up [35] with their own communication protocols before certain organizations began developing open standards. The merging with IIoTS has led to an increase in the number of protocols. In spite of the considerable number of both proprietary and nonproprietary protocols, some are more used than others: Modbus, PROFIBUS (Process Field Bus), Wi-Fi, HTTP (Hypertext Transfer Protocol), CoAP (Constrained Application Protocol), DNP3 (Distributed Network Protocol-3), AMQP (Advanced Message Queuing Protocol), and WiMAX (Worldwide Interoperability for Microwave Access).

From a communication standpoint [36], IoT and CPSs depend heavily on mobile Internet (i.e., telecommunication networks), which haven't played too big of a part in industrial communication thus far. Both IT and telecom networks couldn't manage to meet the requirements for a trustworthy and adept communication. However, things stand differently now as, on the one hand, Ethernet TSN (Time-Sensitive Networking) assures hard real-time capabilities and, on the other hand, the telecom industry sees the industrial automation field as a promising application for their products. Both technological advancements have the chance to alter the structure of industrial networks and even might be a prerequisite for implementing IIoTS and CPSs.

In the early stages of IIoTS communication [36, 37], in order to overpower the constraints brought in by parallel cabling between sensors, actuators, and controllers, as well as to bring to an end the gap on the lower levels of the automation hierarchy, dedicated automation networks named *fieldbus systems* were created. Table 2.1 shows the chronological evolution of industrial communication and notes milestones in diverse technology fields which are relevant to the field of automation.

Up until a few years ago, industrial communication relied on a combination of fieldbus systems, Ethernet-based approaches, and some wireless solutions. But all of these clashed hard with the market. The introduction of IoT and CPS concepts into the automation world put the request of integrating information flows in a wider context: the technological breakthroughs in communication make possible the interconnection on a larger and more fine-grained scale [38].

From an application perspective, the trend is to move the business logic into cloud-based applications. Cloud computing [38, 39] is an element of the system that combines both hardware and software parts in the data centers. There is a series of clouds in this domain: *public cloud*, created as a pay-as-you-go manner for the general public, and *private cloud*, which corresponds to internal data centers of a business that are not available for other organizations or people.

The architecture of the cloud can be divided into four layers: hardware (data center), infrastructures, platform, and application. By themselves, the layers can be seen as consumers for the layer below and as services for the layer above. The cloud platform can be shared into other sections, more accurately in three sections [40]:

- Popular service as a service (SaaS) – it concerns the supplying of applications running in cloud environments. Applications usually can be accessed through a web browser or a thin client.
- Platform as a service (PaaS) – it refers to platform-layer resources (e.g., software development frameworks, operating system support, etc.).
- Infrastructure as a service (IaaS) – it provides processing, storage, and network resources and grants the customer the control of the operating system, the storage, and the applications.

Cloud computing model has the economic advantages of freeing the business owner from the obligation of investing in the infrastructure, renting resources to needs, and paying only for the usage. Furthermore, it favors the decrease in operating costs because of the fact that service providers don't need to supply capacities according to peak load. As far as technical advantages, it ensures certain benefits like energy efficiency, elasticity, and flexibility, optimization of hardware and software resource utilization, and performance isolation [40].

Table 2.1 The evolution of industrial communication and related technology fields

	–1970	1970–1980	1980–1990	1990–2000	2000–2010	2010–2020
Computer networks	**ARPANET** (Advanced Research Agency Network)	**Ethernet** **ISO/OSI** (International Organisation for Standardization)	**Internet** **MAP** (Manufacturing Automation Protocol) **MMS** (Manufaturing Messaging Specification)	**WWW** (World Wide Web) **WLAN** (Wireless Local Area Network)	**Bluetooth** **SOAP** (Simple Object Access Protocol) **ZigBee** **6LoWPAN** (IPV6 over Low – Power WPAN) **UWB** (Ultra Wide Band)	**TSN** (Time-Sensitive Networking)
Mobile internet				**2G: GSM** (Global System for Mobile Communication)	**3G: UMTS** (Universal Mobile Telecommunications System) **3G: HSPA** (High Speed Packet Access) **4G: LTE** (Long-Term Evolution)	**4G: LTE+** **5G**
IT trends			**Ubiq. Comp.**	**IoT** (Internet Of Things)	**CPS** (Cyber Physical System)	**Ind. Internet** **Industry 4.0**
Industrial communication		**Modbus**	**PROWAY** (Process Data Highway) **HART** (Highway Addressable Remote Transducer) **CAN** (Controller Area network) **INTERBUS** **PROFIBUS** (Process Field Bus) **LON** (Local Operating Network) **AS-I** (Actuator Sensor Interface)	**SDS** (Smart Distributed System) **DeviceNet** **ControlNet** **TTP** (Time Triggered Protocol) **LIN** (Local Interconnect Network) **EN50170** **Ethernet/IP** **IEC61158**	**Powerlink** **Modbus/TCP** (Transport Control Protocol) **PROFINET** (Process Field Net) **IEC61784** **EtherCAT** (Ethernet for Control Automation Technology) **OPC UA** **Wireless HART** **IEC61784-2** **ISA 100.11a**	

Data Fusion Based on Advanced AI Algorithms

The data fusion community is in continuous evolution. Applications of data fusion range from DoD (Department of Defense) applications such as automatic target recognition for smart weapons and battlefield surveillance to non-DoD applications such as improved medical diagnosis and condition-based maintenance.

Data fusion was [41] created initially for the military (initiative promoted by DoD). A few applications include automatic target recognition for smart weapons, guidance for autonomous vehicles, or remote sensing. Data fusion implies collecting data from multiple sources (sensors) and combining them for improved performance. It is proven that this technique is better than the standard single source data acquisition. A better example to understand data fusion is the ability of humans and animals to recognize objects. It is not sufficient to tell the quality of an item (e.g., food) just by seeing it, and such, a combination of sight, touch, smell, and even taste provides a much better evaluation of the specific item and suggests a definition of data fusion based on human senses.

In the military domain, an [42] easy example of data fusion would be combining the information obtained from an infrared imaging sensor and a radar. We can use these to determine the exact location of an object, a determination that could not be done by each sensor individually. Different types of sensors are implemented to obtain data fusions, meaning there are many techniques we can use in our system depending on the data that is combined and the nature of the information.

In the military field, there are also included applications such as guidance for autonomous vehicles, target recognition, remote sensing, battlefield surveillance, and automated threat recognition. Military applications have also extended from cognition monitoring of weapons and machinery to monitoring the health status of individual soldiers and assistance to logistics [42].

Surface-to-air and air-to-air defense systems have been developed by the military to detect, track, and identify hostile aircraft, missiles, and antiaircraft weapons. Sensors such as passive electronic support measures, radar, identification-friend-foe sensors, infrared, electro-optic image sensors, and visual (human) sightings are used by these defense systems. Challenges to these data fusion systems include enemy countermeasures, the need for rapid decision-making, and potentially large combinations of target-sensor pairings [41].

These techniques involving data [42] fusion can be performed using advanced AI algorithms (neural networks, genetic algorithms, artificial intelligence), probabilistic approaches (Kalman filter, Bayesian approach with Bayesian network and state-space models), and statistical approaches (multiple variants statistical analysis and its corresponding data mining engine, weighted combination). AI algorithms (Fig. 2.5) are used mainly to derive estimators or classifiers and act as a fusion framework of classifiers/estimators.

If we refer to the nonmilitary field, we have different areas in which data fusion has a significant impact: medical applications, smart buildings, autonomous vehicles, robots, industrial applications, etc. Automatic incident detection can be pre-

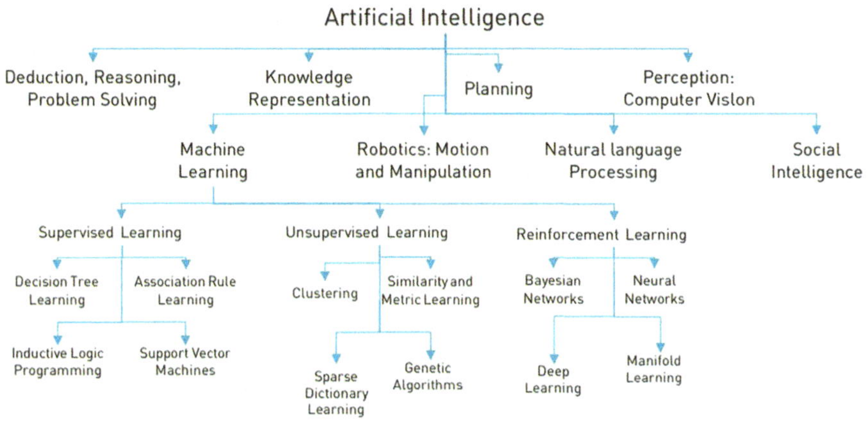

Fig. 2.5 List of AI algorithms [43]

sented as an application of data fusion based on AI algorithms. Koppelman and Ivan developed this application. One of the data sources is a probe vehicle, while the other is a fixed detector (inductive loop detectors). Based on a neural network, two strategies were tested. The first one implies observing the traffic directly and determines whether an incident occurs or not. In the second strategy, different incident detection algorithms preprocess data from each source, resulting in information which is combined by the neural network. It was determined that the latter strategy provided better results and improved performance. Data fusion techniques were also applied in the road network control issues, advanced driver assistance, advanced traveler information systems, and traffic demand estimation [43].

The area of medical diagnosis is another example of data fusion system for non-military applications. Nowadays, in order to provide improvements in medical diagnostic capability, there are being developed increasingly sophisticated sensors for medical applications. The purpose behind fusing these data together is to reduce the chance of false diagnoses and to improve the diagnostic capability [43].

Conclusions

The generation of IIoTS-based platform will favor the creation of a reliable, tangible, and safe open platform which will further allow the integration and development of new services, control software, and applications. In addition, a testing framework for IIoT-based platforms will be developed in order to grant security and reliability to the structure and communication environment so that the data collected from sensors and devices can be exchanged bilaterally and assimilated by other devices and systems.

In the context of production, the evolutionary process will soon get to the point of networked manufacturing systems, which hold a high degree of autonomy.

Systems like these depict autonomous components that come in contact through open networks and semantic descriptions to other field systems, manufacturing modules, and products. Such a thing allows the machines to organize themselves within the network.

Data fusion can connect, reunite, and incorporate data coming from various sources. The scope behind using data fusion is to improve performance by reducing the network traffic and energy consumption, as well as increasing the accuracy of the results. By involving advanced AI algorithms, numerous techniques become this way available, leading to the development of a great number of applications with significant impact.

Acknowledgments This work has been supported in part by UEFISCDI Romania and MCI through projects PARFAIT, OPTIMUM, PAPUD, and Smart-PDM and funded in part by European Union's Horizon 2020 research and innovation program under grant agreement No. 777996 (SealedGRID project), No. 787002 (SAFECARE project), and No. 826452 (Arrowhead Tools).

References

1. Delsing, J. (2017). *IoT automation: Arrowhead framework*. Boca Raton/London/New York: CRC Press.
2. Zuehlke, D. (2018). SmartFactory – Towards a factory-of-things. *Annual Reviews in Control, 34*, 129–138.
3. Gilchrist, A. (2016). *Industry 4.0: The industrial internet of things*. Berkeley: Apress.
4. Wang, K., Wang, Y., Sun, Y., Guo, S., & Wu, J. (2016). Green industrial internet of things architecture: An energy-efficient perspective. *IEEE Communications Magazine, 54*(11), 48–54.
5. Lee, S.-J., Lee, Y.-K., & Yoon, D. (2018). Design considerations for automated crane assignment. *IOP Conference Series: Materials Science and Engineering, 417*(1), 012019.
6. Wang, L., & Wang, G. (2016). Big data in cyber-physical systems, digital manufacturing and Industry 4.0. *International Journal of Engineering and Manufacturing (IJEM), 4*, 1–8.
7. Jeschke, S., Brecher, C., Meisen, T., Özdemir, D., & Eschert, T. (2016). *Industrial internet of things and cyber manufacturing systems* (Springer Series in Wireless Technology) (pp. 3–19). Cham: Springer International Publishing.
8. Lee, Y.-K., Lee, S.-J., & Yoon, D. (2018). *Architecture of middleware system for intelligent factory automation based on industrial internet of things*. International Conference on Big Data, IoT, and Cloud Computing (BIC).
9. Thames, L., & Schaefer, D. (2016). Software-defined cloud manufacturing for Industry 4.0. *Procedia CIRP, 52*, 12–17.
10. Posada, J., Toro, C., Barandiaran, I., Oyarzun, D., Stricker, D., de Amicis, R., & Vallarino, I. (2015). Visual computing as a key enabling technology for Industry 4.0 and industrial internet. *IEEE Computer Graphics and Applications, 35*(2), 26–40.
11. Roblek, V., Meško, M., & Krapež, A. (2016). A complex view of Industry 4.0. *SAGE Open, 6*(2), 1–11.
12. Song, H., Rawat, D., Jeschke, S., & Brecher, C. (2016). *Cyber-physical systems: Foundations, principles and applications*. London: Academic Press.
13. Kagermann, H., Helbig, J., Hellinger, A., & Wahlster, W. (2013). *Recommendations for implementing the strategic initiative INDUSTRIE 4.0: Securing the future of German manufacturing industry*. Final report of the Industrie 4.0 Working Group.

14. Wolf, A. (2017, May 19). Industry 4.0 – Digitisation of manufacturing. *Bosch India Blog* [Online]. Available: https://blog.boschindia.com/iot/industry-4-0-digitisation-of-manufacturing/. Almada-Lobo, F. (2016). The Industry 4.0 revolution and the future of manufacturing execution systems (MES). *Journal of Innovation Management, 3*(4), 16.
15. Lang, S., Reggelin, T., Motasem, J., & Hofmann, W. (2018). *Towards a modular, decentralized and digital Industry 4.0 learning factory*. The 6th International Conference on Enterprise Systems (ES 2018), At Limassol (Cyprus).
16. Scalabre, O. (2018). *Embracing Industry 4.0 – and rediscovering growth*. Boston Consulting Group. Retrieved from https://www.bcg.com/capabilities/operations/embracing-industry-4.0-rediscovering-growth.aspx
17. Weyer, S., Schmitt, M., Ohmer, M., & Gorecky, D. (2015). Towards Industry 4.0 – Standardization as the crucial challenge for highly modular, multi-vendor production systems. *IFAC-PapersOnLine, 48*(3), 579–584.
18. Ivanov, D., Dolgui, A., Sokolov, B., Werner, F., & Ivanova, M. (2016). A dynamic model and an algorithm for short-term supply chain scheduling in the smart factory Industry 4.0. *International Journal of Production Research, 54*(2), 386–402.
19. Stankovic, J.-A., Le, T., Abdeltawab, H., & Tian, Y. (2019). Hardware/software security patches for internet of trillions of things. *arXiv:1903.05266*.
20. Barrett, M.-P. (2019). *Framework for improving critical infrastructure cybersecurity* (Version 1.1).
21. Zhou, L., & Guo, H. (2018). *Anomaly detection methods for IIoT Networks*. IEEE International Conference on Service Operations and Logistics, and Informatics.
22. Sarma, M. (2018). *The global race for artificial intelligence: Weighing benefits and risks*. New Delhi: Institute for Defence Studies & Analyses.
23. Ahemd, M.-M., Shah, M. A., & Wahid, A. (2017). *IoT security: A layered approach for attacks & defenses*. International Conference on Communication Technologies (ComTech).
24. Zhu, C., Rodrigues, J. J. P. C., Leung, V. C. M., Shu, L., & Yang, L. T. (2018). Trust-based communication for the industrial internet of things. *IEEE Communications Magazine, 56*(2), 16–22.
25. Suciu, G., Scheianu, A., Bălăceanu, C. M., Petre, I., Dragu, M., Vochin, M., & Vulpe, A. (2018). Sensors fusion approach using UAVs and body sensors. *Advances in Intelligent Systems and Computing, 747*, 146–153.
26. Wang, M., Perera, C., Jayaraman, P. P., Zhang, M., Strazdins, P., Shyamsundar, R. K., & Ranjan, R. (2016). City data fusion: Sensor data fusion in the internet of things. *International Journal of Distributed Systems and Technologies (IJDST), 7*(1), 15–36.
27. Suciu, G., Suciu, V., Martian, A., Craciunescu, R., Vulpe, A., Marcu, I., Halunga, S., & Fratu, O. (2015). Big data, internet of things and cloud convergence – an architecture for secure E-health applications. *Journal of Medical Systems, 39*(11), 141.
28. Newton, M. (2017). *The state of the industrial internet of things*. https://www.sciencedirect.com/science/article/pii/S0166361517307285.
29. Eden, P., Blyth, A., Jones, K., Soulsby, H., Burnap, P., Cherdantseva, Y., & Stoddart, K. (2017). SCADA system forensic analysis within IIoT. In *Cybersecurity for Industry 4.0* (pp. 73–101). Cham: Springer International Publishing.
30. He, W., Yan, G., & Li Da, X. (2014). Developing vehicular data cloud services in the IoT environment. *IEEE Transactions on Industrial Informatics, 10*(2), 1587–1595.
31. Al-Sarawi, S., Anbar, M., Alieyan, K., & Alzubaidi, M. (2017). *Internet of things (IoT) communication protocols: Review*. 8th International Conference on Information Technology (ICIT).
32. Sajid, A., Abbas, H., & Saleem, K. (2017). Cloud-assisted IoT-based SCADA systems security: A review of the state of the art and future challenges. *IEEE Access, 4*, 1375–1384.
33. Janicke, H., Nicholson, A., Webber, S., & Cau, A. (2015). Runtime-monitoring for industrial control systems. *Electronics, 4*(4), 995.
34. Babu, B., Ijyas, T., Muneer, P., & Varghese, J. (2017). *Security issues in SCADA based industrial control systems*. 2nd International Conference on Anti-Cyber Crimes (ICACC).

35. Wollschlaeger, M., Sauter, T., & Jasperneite, J. (2017). The future of industrial communication: Automation networks in the era of the internet of things and Industry 4.0. *IEEE Industrial Electronics Magazine, 11*(1), 17–27.
36. Zhang, Y., Qian, C., Lv, J., & Liu, Y. (2016). Agent and cyber-physical system based self-organizing and self-adaptive intelligent shopfloor. *IEEE Transactions on Industrial Informatics, 13*(2), 737–747.
37. Armbrust, M., Fox, A., Griffith, R., Joseph, A., Katz, R., Konwinski, A., Lee, G., Patterson, D., Ariel Rabkin, A., Ion Stoica, I., & Zaharia, M. (2010). A view of cloud computing. *Communications of the ACM, 53*(4), 50–58.
38. Botta, A., de Donato, W., Persico, V., & Pescape, A. (2016). Integration of cloud computing and internet of things: A survey. *Future Generation Computer Systems, 56*, 684–700.
39. Xiong, K., & Perros, H. G. (2009). *Service performance and analysis in cloud computing*. 2009 IEEE Congress on Services – I, Los Angeles, CA.
40. Jing, L., Wang, T., Zhao, M., & Wang, P. (2017). An adaptive multi-sensor data fusion method based on deep convolutional neural networks for fault diagnosis planetary gearbox. *Sensors, 17*(2), 414.
41. Faouzi, N.-E. E., Leung, H., & Kurian, A. (2011). Data fusion in intelligent transportation systems: Progress and challenges – A survey. *Information Fusion, 12*(1), 4–10.
42. Il, M. L., Hall, D., & Llinas, J. (2017). *Handbook of multisensor data fusion: Theory and practice* (2nd ed.). Boca Raton: CRC Press.
43. Ahmadi, S. (2014). *LTE-advanced: A practical systems approach to understanding 3GPP LTE releases 10 and 11 radio access technologies*. Waltham: Elsevier.

Chapter 3
Presents the Technology, Protocols, and New Innovations in Industrial Internet of Things (IIoT)

Quanxin Zhao

Innovation and Application

Application field and industrial product are fundamental for IIoT. The primary IIoT application fields include cyber-physical systems, Industry 4.0, machine-to-machine communications, multi-agent systems, and wireless sensor networks. Moreover, the primary industrial products in the applications of IIoT include car and truck, car part, electronic power, electronics, foodstuff, heating, and oil and gas.

Reference architecture is the guidance for designing and producing industrial products. RAMI 4.0, the reference architecture model for Industry 4.0, can be analyzed from the aspects of layers and hierarchy levels. On the one hand, the layers include business, function, information, communication, integration, and asset. On the other hand, the hierarchy levels include connected world, enterprise, work center, station, control device, field device, and product [1–3].

It is heating up to discuss about the Industrial Internet of Things (IIoT) and how it brings benefits to businesses. IIoT can help companies maximize interconnected real-time operational data. IIoT will not only give enterprises quick and flexible decision-making ability, but help companies optimize their operating costs. Making the digital transformation of enterprises possible, industrial IoT technology helps achieve below benefits in a step-by-step manner:

- Automate predictive analysis.
- Better connecting people, assets, and processes.
- Fully optimized enterprise performance.
- Immediately realize the collaborative operation of personnel.

Q. Zhao (✉)
Sichuan Netop Telecom Co., Mianyang, Sichuan, China
e-mail: qxzhao@std.uestc.edu.cn

© Springer Nature Switzerland AG 2020

G. R. Kanagachidambaresan et al. (eds.), *Internet of Things for Industry 4.0*, EAI/Springer Innovations in Communication and Computing,
https://doi.org/10.1007/978-3-030-32530-5_3

Technology and Protocol

The protocol is important for connecting the products. The IIoT communication protocols can be classified into some kinds, which are as follows: First, the wireless sensor network includes WirelessHART, IEC 62591 (WirelessHART), ISA 100.11a, and Zigbee. Second, the M2M communication includes CoAP, OPC-UA, DDS, and Modbus. Third, the messaging includes MQTT, AMQP, and XMPP. Fourth, the low-power wide-area network (LPWAN) includes NB-IoT, Sigfox, LoRa, and LoRaWAN. Fifth, the cellular network includes 5G, 4G, 3G, LTE, WiMax, GPRS, and GMS. Sixth, the wireless local area network (WLAN) includes IEEE 802.11. Seventh, the wireless personal area network (WPAN) includes IEEE 802.15.4. Above protocols are described below [4–6].

WLAN

Wireless local area network (WLAN), a data transmission system, which uses the technology of radio frequency (RF) to substitute the existing local network of twisted copper wires (i.e., coaxial) by electromagnetic waves to make the LAN with wirelessness. With a simple access architecture, the network enables users to achieve the ideal realm of "information and convenience."

WPAN

The communication technology of wireless personal area network (WPAN) utilizes wireless communication connection. Recommended by IEEE, there exist three technologies, which are 802.15 on the basis of Bluetooth technology, high-frequency 802.15.3 (i.e., UWB), and low-frequency 802.15.4 (i.e., Zigbee). WPAN is designed to achieve activities within tiny range, business emerging wireless communication network technology with rich types, specific groups, and wireless seamless connections [7–10].

5G

In the technology of mobile cellular communication, the fifth-generation mobile communication technology (i.e., 5G) is the up-to-date. Compared with previous mobile cellular communication, such as 4G (LTE-A, WiMax), 3G (UMTS, LTE), and 2G (GSM), there exists significant change in the performance, such as throughput and system capacity, jitter and latency, lower power consumption and costs, and massive attached devices with massive connectivity [11–15].

There are two phases for 5G. In Release-15, the first phase is to accommodate deployments of existing commercial. In Release-16, the second phase can be a candidate technology of IMT-2020 for the International Telecommunication Union (ITU). In the IMT-2020 specification of ITU, its speed can be up to 20 Gbit/s contributed by both high-capacity Multi-Input-Multi-Output (MIMO) and large channel bandwidth.

1G/2G/3G/4G

In 1978, the first large-capacity mobile cellular communication system in the world, named the Advanced Mobile Phone Service (AMPS) system, is designed by Bell Labs. In order to access the public telephone network automatically for the mobile terminals, frequency reuse technology is adopted by AMPS within the entire service coverage area of the system. Therefore, it has better voice quality and greater system capacity and further solves the large capacity faced by the public mobile communication system. It requires conflicts with spectrum resource limitations. The beginning to deploy on a large-scale deployment of the AMPS systems is started by the United States in the late 1970s. AMPS has won unanimous praise from users for its excellent network performance and service quality, which has facilitated on the global scale of researching cellular mobile communication technologies starting after AMPS's fast development in the United States.

The ETACS system, the NTT/JTACS/NTACS system, and the NMT-450 system are developed by Europe and Japan by the mid-1980s. These cellular mobile communication networks are analog-based frequency division duplex (FDD) systems (known as 1G systems) [16–20].

900/1800 MHz GSM Mobile Communication

GSM mobile communication network (known as 2G), working in the 900/1800 MHz frequency band, can offer voice and data services. In the GSM mobile communication system, the TDMA technology is adopted by the wireless interface, and the MAP protocol is adopted by the core network mobility management.

The 900/1800 MHz GSM second-generation mobile cellular communication service includes types of services:

- End-to-end double-way voice services
- Mobile short-message service, using the mobile station initiated by the GSM network and the message platform, and the message service received by the mobile station
- Mobile data services
- Mobile bearer services
- Mobile supplementary services, for example, call forwarding service, display of calling numbers

- Mobile intelligent network services, for example, prepaid services
- Domestic roaming and international roaming services

Operators which have the 900/1800 MHz GSM second-generation mobile cellular communication service establish their GSM networks, and the provided type of mobile communication service may be part or all. A mobile communication service is provided by a network of different operators. It is necessary to establish state approved international communication gateways for mobile network accessing services of international communications.

800 MHz CDMA Mobile Communication

800 MHz CDMA mobile communication, operating in the 800 MHz frequency band, is also the second-generation mobile cellular communication service providing the voice and data services. Narrowband code division multiple access (CDMA) technology is adopted in wireless interface. CDMA, as well as IS-41 protocol, is adopted in the core network mobility management.

The CDMA working in 800 MHz acts as the second-generation mobile cellular communication service which includes types of services:

- End-to-end double-way voice
- Mobile message service, using the mobile station initiated by the CDMA network and the message platform, and the message service for mobile stations
- Mobile bearer services
- Mobile data services
- Mobile supplementary services, for example, call forwarding service and display for calling number
- Mobile intelligent network services, for example, prepaid services
- Domestic roaming services
- International roaming services

Operators of the 800 MHz CDMA second-generation mobile cellular communication service should establish their mobile CDMA communication networks; besides, the provided type of mobile communication service may be part or all. The network through which a mobile communication service is provided may be the same operator's network or may be completed by a network of different operators.

The third-generation mobile cellular communication (i.e., 3G) service provides voice services, data services, and video image services.

3G mobile communication services provide services of broadband multimedia, including 144 kbps in high-speed mobile circumstance, 384 kbps in walking/slow-moving circumstance, and 2 Mbps data in indoor circumstance. Transmission and guarantee high reliability of service quality (QoS). The third generation of mobile cellular communication services includes all kinds of mobile services including multimedia.

The operator of the 3G mobile communication service must establish a 3G mobile communication network by itself, and the type of mobile communication service provided may be part or all of the end-to-end service. The network through which a mobile communication service is provided may be the same operator network facility or may be completed by network facilities of different operators. International communication services must be approved and established by the state for international communication services.

Although the 3G system had solved the shortcomings of 1G and 2G systems, its actual speed was far from the expected value. Then the international organizations 3GPP and 3GPP2 started a new round of 3G evolution plan, and LTE stood out among many candidate standards. The 3GPP organization launched the "LTE Plan," which achieves a smooth transition from 3G to 4G, so LTE is also known as the quasi-4G standard. The ultimate goal of the program is to provide a low-latency, high-throughput, large-scale coverage wireless communication network. LTE has two working modes: FDD and TDD. Among them, LTE-TDD has its own intellectual property rights. It was commercialized in China at the end of 2013, and its high-speed bandwidth capability has brought a new experience to users. Currently, more than 4 million LTE base stations have been deployed worldwide, and this number is expected to grow with future growth.

NB-IoT

In Internet of Everything, NB-IoT (i.e., the Narrowband Internet of Things) is one of the important areas. NB-IoT developed on cellular network technology usually work together with UMTS networks, GSM networks, or LTE networks consuming 180 kHz of bandwidth.

For the emerging technology of low-power wide-area network (LPWAN), NB-IoT provides wireless data connection in WAN for massive low-power devices. For those low-power devices, they have the characteristics of long standby time and coverage of indoor wireless data connection.

XMPP

XML is the standard universal markup language. As a subset environment of XML, XMPP protocol has flexible to development and scalable application. To develop applications or add new functionality to existing customer systems easily, software protocol included in XMPP on the server side enables the server talking to another one.

XMPP defines roles of client, server, and gateway. The server has following functionalities, such as recording the client information, connecting management, and routing information. To connect heterogeneous instant messaging systems, the gateway is used.

AMQP

As open standard advanced message queuing protocol in an application layer, the Advanced Message Queuing Protocol (AMQP) can give services of unified messaging for message-oriented middleware. Based on AMQP, clients and message middleware transmit messages, no matter the client/middleware different products, different development languages, and so on. Implementations in Erlang include RabbitMQ and more. It supports the architecture of various message exchanges: store and forward, distributed transactions, publish subscriptions, content-based routing, file transfer queue, and point-to-point connection.

MQTT

Message Queuing Telemetry Transport (MQTT) is designed on the basis of paradigm of the publish/subscribe according to ISO/IEC PRF 20922. MQTT, a publish/subscribe messaging protocol, is running on the TCP/IP protocol suite, which is suitable for low-end hardware remote devices with conditions of poor network connection. A message middleware is required to do this.

The MQTT has features as follows:

1. One-to-many message publishing is applied by publish/subscribe message mode.
2. Payload content shield during message transmission.
3. TCP/IP network connection provision.
4. Message publishing service classified for three kinds:
   - "At most once" where message loss/duplication is permitted, as which depends on the underlying TCP/IP network.
   - "At least once" to permit message duplication for ensuring the arrival of message.
   - "Only once" ensuring the arrival of message only once.
5. Small transmission, low overhead, and network traffic are reduced to minimum during protocol exchange.

Modbus

For communication with programmable logic controllers (PLCs), a serial communication protocol (Modbus) is published in 1979. In the area of industrial communication protocols (De facto), the industry standard Modbus can connect industrial electronic devices. The main reasons why Modbus is used more widely than other communication protocols are:

1. Published and no copyright requirements.
2. Easy to deploy and maintain.
3. There are not many restrictions for the vendor to modify the mobile local bits or bytes.

Modbus allows multiple (approximately 240) devices to be connected to communicate on the same network, for example, a device that measures temperature and humidity and sends the results to a computer. In the Data Acquisition and Monitoring Control System (SCADA), Modbus is typically used to connect monitoring computers to remote terminal control systems (RTUs).

DDS

DDS (Data Distribution Service) is a distributed real-time data distribution service middleware protocol, which is "TCP/IP" in distributed real-time network and used to solve network protocol interconnection in real-time network, whose function is equivalent to "bus on bus":

What is a distributed real-time network system?

For a real-time system, it focuses on the correctness of calculation, which usually relies on the result of correctness of program logical and time. There will be system error if system time constraints are not met. Therefore, it requires the system to timely respond to external events requests, to finish the event processing with time constraint, and timely control and coordinate tasks operation.

The system is the normal state of reality, from the defense system to the temperature and humidity remote control system of the warehouse, which is common in all walks of life.

OPC-UA

The core of the OPC Unified Architecture (OPC UA) communication standard is interoperability and standardization. The traditional OPC technology solves the problem of interoperability between hardware devices at the level of control; besides, communication standardization at the level of enterprise is also required. Prior to OPC UA, the access specifications were based on Microsoft's COM/DCOM technology, which would give new levels of communication a weakness that cannot be eradicated. In addition to the inflexibility of traditional OPC technology and the limitations of platforms, the OPC Foundation has released the latest unified method of data communication, covering timely OPC data access specifications, different aspects of the OPC Historical Data Access Specification, the OPC Alarm Event Access Specification, and the OPC Security Protocol, but with functional extensions.

OPC UA is another breakthrough after the success of traditional OPC technology, making data collection, information modeling, and communication reliable and secure.

CoAP

In the CoAP network, there are massive resource-constrained devices; therefore, devices with small space of memory and power with limited value cannot afford traditional large HTTP protocol application. Then, a CoAP protocol is proposed on the basis of the REST architecture by the IETF's CoRE working group.

In the 6LowPAN protocol stack, CoAP protocol is designed on the layer of application. After describing the content, features, and interaction model of the CoAP protocol in detail, this article uses the Contiki embedded operating system on the uIPv6 START KIT wireless network development kit, which both implements browser a CoAP protocol and uses its own client. The CoAP protocol implemented program increases function of database interaction, thereby realizing web interface the function of real-time and historical data view.

Zigbee

Zigbee is a low-speed short-distance communication protocol in wireless networks. In the layers of the media access and physical, Zigbee applied the IEEE 802.15.4 standard, which has the features of low speed and power consumption, low cost and complexity, reliable and secure, supporting massive online nodes, and multiple online topologies.

ISA100.11a

ISA100.11a is one of the international standards for industrial wireless sensor networks. On the basis of IEEE 802.15.4, but only using its 2.4 GHz ISM band (without sub-1 GHz band). It has been approved by the International Electrotechnical Commission (IEC) in September 2014 and has become the official international standard; the standard number is IEC 62734.

The protocol defines the following five types of device roles:

1. The upper computer control system is a platform for users and engineers to interact with the ISA100.11a system.
2. Gateway defines host computer interfaces to the network and is also the interface with other factory-level networks. Multiple gateways can exist in an ISA100.11a network system.

3. The backbone router (BBR) is the infrastructure of the ISA100.11a backbone network and can provide data routing in the network of backbone. The communication protocol on the ISA100.11a backbone network can be a wireless protocol, such as Wi-Fi, or a wired protocol, such as a standard Ethernet. That is, the backbone of ISA100.11a is another high-performance network.
4. Field devices, which have one terminal node device and two field routers. Terminal node devices typically have sensors/actuators. In addition to the sensor/actuator, the field router also has a routing function to route data of the terminal node device within the ISA100.11a subnet.
5. Handheld device is a device that accesses the ISA100.11a system for on-site maintenance and configuration.

WirelessHART

WirelessHART is designed for the first wireless communication standard with open interoperable characteristics to satisfying process industry plant applications critical needs such as stable, secure, and reliable during wireless communications.

Each WirelessHART network has the following components:

1. Field device attached to the plant equipment or process.
2. Gateways enabling wireless field devices to communicate with host applications with the connection of the backplane and plant-level communication networks.
3. Network management software is to configure network, schedule devices communication, manage packet routing, and monitor health of network.

Standards

Standard and stack can guide the company for developing products. The IIoT communication standard and technology stack can be classified into layers and groups. On the one hand, layers are consisting of a framework layer, a transport layer, a network layer, a link layer, and a physical layer. On the other hand, main groups include WPAN, WLAN, cellular network, LPWAN, satellite network, and traditional industrial computer network (fieldbus), which will be described in detail.

Fieldbus

Fieldbus was developed internationally in the late 1980s and early 1990s for field intelligent device interconnection communication networks in the fields of process automation, manufacturing automation, and building automation. Fieldbus serves as the basis for the plant's digital communication network and communicates the links between the production process site and control equipment with high control and management level. Fieldbus is a fully control system as well as a distributed grassroots network. This integrated technology with intelligent sensing, control, computer, digital communication, and other technologies has attracted worldwide attention and will lead to profound changes in the structure and equipment of automation systems. Many internationally powerful and influential companies have developed fieldbus technologies and products from varying degrees.

The fieldbus system (known as the fieldbus control system (FCS)) is the 15th-generation control system. People refer to the FCS as the first generation of the Pneumatic Signal Control System before the 1950s. The electric analog signal control system such as 4–20 mA is called the second generation, and the digital computer centralized control system is called the third generation. The distributed control system (DCS) since the middle of the decade is called the fourth generation The FCS breaks through the limitations of the DCS using a communication-dedicated network and adopts an open and standardized solution to overcome the closed system defects. Meanwhile, the centralized and decentralized system of distributed systems has become fully distributed structures, and the functions of control have been fully decentralized to the scene. It can be said that fieldbus system prominent features are openness, digital, and dispersion communication.

Fieldbus technology has gone through the initial stages of fragmentation and separation, and the current standard is IEC 61158. Among them, there are strong strengths and influences: Foundation Fieldbus (FF), LonWorks, Profibus, HART, CAN, Dupline, etc. They have their own characteristics and have formed their own advantages in different application areas. Here, it will briefly describe the characteristics of fieldbus technology, closely related to the reliability and practicability of the system, and introduce the key technologies such as fieldbus network structure and architecture and the current status of several popular fieldbus technologies. Finally, the development trend and technology prospect of fieldbus are explained.

First, the technical characteristics of the fieldbus are as below:

1. The openness of the system. An open system means that the communication protocol is open, and devices of different manufacturers can be interconnected and exchange information. Fieldbus developers are committed to establishing an open system of a unified factory underlying network. For standards, openness here refers to the consistency and openness. A fieldbus network open system gives users the right to integrate the system. Users can make products from different suppliers into random systems according to their own needs and objects.
2. Interoperability and interoperability. Interoperability here refers to the realization of information transmission and communication between interconnected devices and systems. It supports digital communication of point to point and point to multipoint. Interoperability refers to devices from different manufacturers with similar performance support interchanging for interoperability.
3. Functional autonomy and intelligent field devices. It distributes the functions of sensing measurement, compensation calculation, and processing and controlling engineering quantity. The automatic control functions are completed only by the field device and diagnosing equipment operation status.

4. Highly dispersed system structure. Since the field device itself can complete the basic functions of automatic control, the fieldbus has formed a new architecture of a fully distributed control system. It fundamentally changed the system of distributed control system combining the concentration and dispersion of existing DCS, improving reliability, and simplifying system structure.
5. Adaptability to the on-site environment. Working at the front end of the field equipment, as the fieldbus on the bottom of the factory network, designing for the environment of field. The capability of interference can be realized by two-wire system for power transmitting and communicating for satisfying requirements of intrinsic safety.

Second, the advantages of fieldbus are below.

Due to the above characteristics of the fieldbus, especially the simplification of the structure of the fieldbus system, the design, installation, and commissioning of the control system to the normal production operation and maintenance and repair thereof all show superiority:

1. Save hardware quantity and investment. Because the smart devices scattered in the front end of the fieldbus system device process kinds of functions such as alarm, control, calculation, and sensing. Therefore, transmitter number can be reduced, eliminating the need for separate controllers, computing units, etc. In addition, it is required for DCS with the technology of signal conversion, isolation and conditioning, and their complex wiring. Industrial PCs can also be used as operating stations, with savage investment of hardware. As reducing control equipment, the control room can be reduced.
2. Save installation costs. The wiring of the fieldbus system is very simple. Because a pair of twisted pairs or a cable can usually be connected to multiple devices, the amount of cables, terminals, slots, and bridges is greatly reduced. The wiring design and the proofreading work of the joints are also decreased very much. When it is necessary to add on-site control equipment, it is not necessary to add a new cable, and it can be connected to the original cable nearby, which saves investment and reduces the workload of design and installation. According to the calculation data of the typical test project, the installation cost can be saved by more than 60%.
3. Save maintenance overhead. Because the field control equipment has the ability of self-diagnosis and simple fault handling and through the digital communication to send relevant diagnostic maintenance information to the control room, the user can query the operation of all equipment and diagnose the maintenance information, so as to analyze the cause of the fault early and quickly eliminate it. Maintenance downtime is reduced, and maintenance is reduced due to simplified system structure and simple wiring.
4. The user has a high degree of system integration initiative. Users are free to choose the devices provided by different vendors to integrate the system. Avoiding the choice of a certain brand of products is "framed dead" equipment; it will not be incompatible with the incompatible protocols and interfaces in the system integration, so that the initiative in the system integration process is completely in the hands of users.

5. Improve system accuracy and reliability. Due to the intelligent and digitized fieldbus devices, it fundamentally improves measurement and control's accuracy and reduces transmission errors compared with analog signals. At the same time, due to system structure which is simplified, it reduces equipment as well as wiring and strengthens internal field instrument functions: the round-trip signal transmission is reduced, and the operational reliability of the system is improved. In addition, due to its equipment standardization and functional modularization, it also has the characteristics such as easy reconstruction as well as simple design.

AMPS

The AMPS (Advanced Mobile Phone System), as well as NMT and NTT, is one of typical representatives of mobile communication systems in the United States. AMPS's analog cellular transmission works on the 800 MHz band which is used in America, as well as parts of the Pacific Rim.

From the mid-1970s to the mid-1980s, the first generation was a simulated mobile cellular communication network. Compared to previous mobile communication systems, the breakthrough was that Bell Labs proposed cellular networks concept by in the 1970s. The cellular network, that is, the cell system, greatly increases the system capacity due to frequency reuse.

Total Access Communication System (TACS), which is divided into the version of both ETACS (Europe) and NTACS (Japan), occupies the 900 MHz band. This standard is widely used in some Asian countries, Japan, and the United Kingdom.

The first-generation mobile communication system has the main feature which is the frequency division multiplexing. Moreover, analog modulation is applied for the voice signal, with 30/25 kHz used for and each analog user channel. Great commercial success has been achieved by the first-generation system; however, it leaves some drawbacks in the areas of:

1. Spectrum utilization
2. Business types
3. Data service speed
4. Confidentiality
5. Cost of equipment
6. Size and weight

As the analog system has such fundamental technical defects, people further developed and emerged digital mobile communication technology (i.e., the second-generation mobile communication system) from the mid-1980s. First, the Pan-European Digital Mobile Telecommunications Network (GSM) system was first introduced by Europe. Next, Japan and the United States designed their mobile digital communication systems. Compared with analog mobile communication, digital mobile communication network improves utilization of spectrum and supports many services. The second-generation mobile communication system (i.e., narrowband digital communication system) aims to transmit low-speed data and voice services. In the second generation of digital cellular mobile communication systems, typical representatives are the IS-95, US DAMPS, and the European GSM system:

1. GSM is designed for global digital cellular communications, which is a DMA standard and originated in Europe. Interconnected with ISDN, it supports 64 kbps data rate. DCS1800 used by GSM represents the 900 MHz band and the 1800 MHz band. FDD duplex and TDMA multiple-access modes are adopted by GMS, supporting eight channels per carrier frequency with 200 kHz signal bandwidth. The is relatively complete, and the technology is relatively mature. The shortcoming of GSM standard system is analog system capacity which is limited and cannot be further increased.
2. Digital Advanced Mobile Phone System (DAMPS, known as IS-54) also uses the 800 MHz band with TDMA multiple access, which is one of the two early introductions of the digital cellular standard of North America.
3. Another North American digital cellular standard, IS-95, uses the 800/1900 MHz frequency band with CDMA multiple access.

Since the second generation of mobile communication aims to transmit low-speed data and voice services, since 1996, there have been 2.5 generations of mobile communication systems (i.e., IS-95B and GPRS) for the medium-speed data transmission problem. The main mobile communication services currently are still services of low-rate data and voice. Due to network development, the multimedia and data communication develops very fast. So, broadband multimedia mobile communication is becoming the goal for the third-generation mobile communication. From the perspective of development, in the third-generation mobile communication, the core technology is CDMA due to its own technological advantages. The following requirements are imposed on 3G wireless transmission technology (RTT, radio transmission technology):

1. High-speed supporting services of multimedia transmission. The environment of indoor has above 2 Mbps transmission data rate. The indoor and outdoor walking environment is at least 384 kbps. The outdoor vehicle movement is at least 144 kbps, and the satellite mobile environment is at least 9.6 kbps.
2. The transmission rate can be allocated as needed.
3. The links of uplink and downlink can adapt to the asymmetric demand.

In 1985, the International Telecommunication Union (ITU) proposed the third-generation mobile communication system, which was named FPLMTS (the Future Public Land Mobile Telecommunication System) and renamed IMT-2000 (International Mobile Telecommunication-2000) in 1996. It is because the system achieves maximum service rate up to 2000 kbps operating at the 2000 MHz frequency band. It is expected to be commercially available around the year 2000. The main systems are cdma2000, WCDMA, and TD-SCDMA. In 1999, ITU-R TG8 18th meeting adopted the "IMT-2000 Radio Interface Technical Specification" proposal.

WLAN

It can be easily networked, because WLAN can easily and quickly accept new employees without having to make too many changes to the user management configuration of the network. WLAN is easier to implement in places where wired network cabling is difficult. The WLAN solution eliminates the need for perforated wiring and doesn't cause any damage to the building.

For every user using a WLAN, WLAN performance is extremely important. But under some special cases, it is especially important than other features, so it has to talk about this first. Regarding WLAN performance, we usually refer to the optimal transmit and receive throughput of a WLAN over a wireless signal. The IEEE standard always provides the theoretical throughput limit for each 802.11 protocol. For example, 802.11g has a theoretical maximum physical rate of 54 Mbps. In comparison, the latest second-wave 802.11ac protocol can achieve a maximum theoretical rate of 2.34 Gbps. But it is worth noting that these bit rates have never been true.

The distance of the wireless access point, the number of terminals connected to the AP, wireless signal congestion, and physical obstacles all greatly reduce the actual throughput.

To minimize performance loss, WLAN vendors use a variety of techniques to improve wireless chips and antennas. If your organization requires very high performance for a particular application or if your deployment environment has a significant disruption that can severely degrade the performance of an enterprise WLAN, then you need to choose a vendor that provides the best WLAN performance.

In 1990, the IEEE 802.11 WLAN standard working group is established by the IEEE 802 Standardization Committee. IEEE 802.11 (i.e., Wi-Fi) is an approved standard in June 1997 that designs the physical layer as well as media access control layer specifications. The physical layer designs the characteristics of signal as well as data transmission modulation. Further, it defines two methods of RF transmission as well as one method of infrared transmission. The standard of RF transmission uses direct sequence spread spectrum as well as frequency-hopping spread spectrum and operates in the frequency band of 2.4000–2.4835 GHz. IEEE first developed a wireless LAN standard named IEEE 802.11. IEEE 802.11 is applied solving user wireless access in campus networks as well as in office. The service rate reaches only 2 Mbps. IEEE 802.11b replaces the IEEE 802.11 standard because 802.11 cannot have good transmission distance and speed performance.

In 1999, the officially approved IEEE 802.11b specifies the WLAN with operating 2.4–2.4835 GHz frequency band with 11 Mbps transmission rate and 50–150 ft transmission distance. This standard can supplement to IEEE 802.11. It adopts coded modulation mode with two modes of operation: modes of basic and point to point. For the rate of data transmission, it can be at different rates of 11, 5.5, 2, and 1 Mbps according to actual conditions. Automatic switching changes the design of WLAN and expands WLAN application areas. IEEE 802.11b adopted by many manufacturers has becoming WLAN standard mainstream. The widely used products are introduced in airports, hotels, homes, offices, stations, etc., industry concerned IEEE 802.11a as well as IEEE 802.11g due to many latest WLAN standards emergency.

It developed the IEEE 802.11a standard in 1999, which specifies the frequency band of WLAN operating on 5.15–5.825 GHz, with 54/72 Mbps (Turbo) data transmission rate, and 10–100 m controlled transmission distance. This standard supplements IEEE 802.11, which expands the standard layer of physical, uses technology of OFDM's spread spectrum with QFSK modulation, and provides 10 Mbps wired Ethernet and 25 Mbps wireless ATM. The interface of wireless frame structure supports many services including data, images, and voice, and one sector can support many mobile terminals and any terminal. The standard follow-up IEEE 802.11b is IEEE 802.11a. The original design of IEEE 802.11a aims at substituting 802.11b. However, operating in 2.4 GHz band requires no license. This band belongs to the special frequency bands of industry, education, medical, etc. and is public. It requires a license to operate in the 5.15–8.825 GHz band.

At present, IEEE has introduced the latest version of the IEEE 802.11g authentication standard, which proposes a transmission rate of IEEE 802.11a, which is better than IEEE 802.11b. It adopts two modulation modes, including OFDM and IEEE 802.11b used in 802.11a. CCK is compatible with both 802.11a and 802.11b. It is more suitable to deploy 802.11a for WLAN operators; WLAN enterprises seem very likely choosing 802.11g in order to balance the investment in existing 802.11b devices.

The IEEE 802.11i standard has authentication of device as well as user port in IEEE 802.1x to modify and integrate the WLAN MAC layer and defines a mechanism of authentication with encryption format for improving security of WLAN. The IEEE 802.11i new revised standard mainly includes "technology of Wi-Fi Protected Access (WPA)" as well as "Strong Security Network (SSN)." In 2004, the Wi-Fi Alliance adopts the 802.11i standard as the second version of WPA and began implementation. It is important to construct WLAN with the IEEE 802.11i standard. The data security is the first priority that WLAN equipment manufacturers and WLAN operators should consider first.

The IEEE 802.11e standard proposes improvements to the MAC layer protocol of WLAN supporting multimedia data transmission to support the quality of service (QOS) guarantee mechanism for all WLAN radio broadcast interfaces. IEEE 802.11f, defining access nodes' communication, supports IEEE 802.11 Access Point Interoperability Protocol (IAPP). IEEE 802.11h is for managing spectrum of 802.11a.

IEEE 802.15.4

Specifications, such as Wi-Fi, WirelessHART, and ZigBee, have the basis of IEEE 802.15.4. It defines low-rate wireless personal area networks' layer of physical and protocol of media access control. The rate of data transmission will be up to 250 kbps in the ISM band of both 868/915 MHz and 2.4 GHz.

WPAN

With the communication technology developing rapidly, there exists the requirement for communicating in meters around person's itself. Therefore, the concept of wireless personal area network (WPAN) as well as personal area network (PAN) is proposed. The WPAN can help wireless devices within a close geographical area to set up wireless connections to communication and connect each other. In 2002, it established the IEEE 802.15 working group, which concentrates on the WPAN standardization in the area of the physical layer (PHY) as well as media access layer (MAC). Wireless devices provide communication standards.

In IEEE 802.15 working group, different standards of applications are developed by four different task groups (TGs). These standards of applications are different from the following aspects, such as consuming power, services provision, and transmitting rate. The primary targets of above task groups are as below:

1. Task group TG1 focuses on developing the standard for Bluetooth wireless personal area network, numbered IEEE 802.15.1. This standard concentrates on short-range communication and proposes a close-range, medium-speed WPAN.
2. Task group TG2 focuses on developing coexistence solution between IEEE 802.11 and IEEE 802.15.1, numbered IEEE 802.15.2.
3. Task group TG3 focuses on developing multimedia applications in the wireless personal area network for achieving high data rate as well as quality of service, numbered IEEE 802.15.3.
4. Task group TG4 focuses on developing low-cost, low-rate transmission and uniform standard for low-speed interconnection for low-rate wireless personal area network (LR-WPAN), numbered IEEE 802.15.4.

The LR-WPAN is a wireless communication network with low-cost, enabling applications under wireless connections with low energy and low cost. In comparison with WLAN, LR-WPAN require no infrastructures. For the LR-WPAN, the PHY layer as well as MAC sublayer protocols are defined in IEEE 802.15.4.

In IEEE 802.15.4 standard, the LR-WPAN should be of the characteristics as below:

1. Different carrier frequencies give diverse transmission rates of 20, 40, and 250 kbps.

2. Support star as well as peer-to-peer network topologies.
3. There are two address formats of 16-bit as well as 64-bit.
4. CSMA-CA (carrier sense multiple access with collision avoidance).
5. Support mechanism of acknowledgment (ACK) ensuring the reliability of transmission.

Conclusion

Along the life cycle, product with embedded smart electronics can definitely form an internal even global information network, making IoT with industrialization. Such trend has made the vision of Industry 4.0 closer to the reality, i.e., a fully context-sensitive, controllable, describable, manageable, self-regulating manufacturing system.

There can be drawn a conclusion that IIoT has mandatory requirement for Industry 4.0, which is because IIoT is an emerging approach concentrating on the area of "Computer Networks and Communications" for delivering "a Comprehensive Broadband Infrastructure."

Giving the reader a systematical view of IIoT including the innovation and application, the technology, the protocol, and standards, this chapter shows that smart devices are changing in the world, in which significant trend has already been extended from people's daily life to the world industry.

In the upcoming Industry 4.0, the connected smart devices all around the world via the Internet will provide secure, real-time, and reliable services of sensing, communicating, and computing, making effects on the following aspects, such as chips, terminals, base stations, networks, software tools, testing devices, operating systems, and APPs.

References

1. Alcaraz, C. (2019). *Security and privacy trends in the industrial internet of things*. Cham: Springer.
2. Wan, J., Lai, C. F., Song, H., et al. (2019). Software-defined industrial internet of things. *Wireless Communications and Mobile Computing, 2019*, 1–2.
3. Aazam, M., Harras, K. A., & Zeadally, S. (2019). Fog computing for 5G tactile industrial internet of things: QoE-aware resource allocation model. *IEEE Transactions on Industrial Informatics, 15*(5), 3085–3092.
4. Matthyssens, P. (2019). Reconceptualizing value innovation for Industry 4.0 and the industrial internet of things. *Journal of Business & Industrial Marketing*. https://doi.org/10.1108/jbim-11-2018-0348.
5. Vitturi, S., Sauter, T., & Pang, Z. (2019). Real-time networks and protocols for factory automation and process control systems. *Proceedings of the IEEE, 107*(6), 939–943.
6. Jaloudi, S. (2019). Communication protocols of an industrial internet of things environment: A comparative study. *Future Internet, 11*(3), 66.

7. Salonikias, S., Gouglidis, A., Mavridis, I., et al. (2019). Access control in the industrial internet of things. In *Security and privacy trends in the industrial internet of things* (pp. 95–114). Cham: Springer.
8. Ludwig, S., Karrenbauer, M., Fellan, A., et al. (2018). A5G architecture for the factory of the future. In *2018 IEEE 23rd International Conference on Emerging Technologies and Factory Automation (ETFA)* (Vol. 1). IEEE, pp. 1409–1416.
9. Liao, Y., Loures, E. F. R., & Deschamps, F. (2018). Industrial internet of things: A systematic literature review and insights. *IEEE Internet of Things Journal, 5*(6), 4515–4525.
10. Unverdorben, S., Böhm, B., & Lüder, A. (2018). Reference architectures for future production systems in the field of discrete manufacturing. In *2018 IEEE 14th International Conference on Automation Science and Engineering (CASE)*. IEEE, pp. 869–874.
11. Zhang, Y., & Chang, J. (2016). Application of fieldbus control system on AC/DC microgrid. In *Advances in energy, environment and materials science: Proceedings of the International Conference on Energy, Environment and Materials Science (EEMS 2015), Guanghzou, PR China, August 25–26, 2015* (p. 15). Boca Raton: CRC Press.
12. Huang, B. W., & Li, J. Y. (2014). Design of digital teaching platform of the internet of things. In *Applied mechanics and materials* (Vol. 670, pp. 1671–1674). Zug: Trans Tech Publications.
13. Liu, P., Liu, X. P., & Wang, B. P. (2012). Application of the measurement and control network on rewinder. In *Advanced materials research* (Vol. 463, pp. 1509–1512). Stafa-Zurich: Trans Tech Publications.
14. Liang, Q. (2012). *Communications, Signal Processing, and Systems*. Proceedings of the International Conference on Communications, Signal and Processing, Springer.
15. Ho, Q. D., Gao, Y., Rajalingham, G., et al. (2014). *Wireless communications networks for the smart grid*. Berlin: Springer.
16. Li, Z. D., Feng, G. H., Feng, X., et al. (2015). Design of gas density monitor system based on wireless sensor network. In *Applied mechanics and materials* (Vol. 716, pp. 888–891). Zug: Trans Tech Publications.
17. Wu, X., & Tang, D. (2009). Hierarchical software architecture of real-time monitoring and management system for area water resources based on OPC. In *2009 International Conference on Information Technology and Computer Science* (Vol. 2). IEEE, pp. 287–291.
18. Grangel-González, I., Halilaj, L., & Auer, S., et al. (2016). An RDF-based approach for implementing Industry 4.0 components with administration shells. In *2016 IEEE 21st International Conference on Emerging Technologies and Factory Automation (ETFA)*. IEEE, pp. 1–8.
19. Hankel, M., & Rexroth, B. (2015). *Industrie 4.0: The reference architectural model Industrie 4.0 (RAMI 4.0)*. ZVEI: Die Elektroindustrie.
20. Kagermann, H., Helbig, J., & Hellinger, A., et al. (2013). Recommendations for implementing the strategic initiative INDUSTRIE 4.0: Securing the future of German manufacturing industry; final report of the Industrie 4.0 Working Group. Forschungsunion.

Chapter 4
Decision Making Models Through AI for Internet of Things

E. P. Ephzibah, S. Sree Dharinya, and L. Remya

Biologically inspired computation had taken its first step some decades ago. Still there is a need for optimization of many problems in real-time scenarios. When found in nature, there are algorithms available that have more possibilities of optimization and can help humans to solve naturally hard problems using limited time and space. Some of those techniques are discussed in this chapter as they provide valuable and time-conserving methods that are observed from the nature. This particular area of study binds the subareas like connectivity, communication, and social performance with emergence and existence. This field can be well flourished with the help of artificial intelligence (AI) and machine learning as it involves training combined with decision-making for optimization. This field of study requires knowledge in biology as it monitors the behavioral aspects of the insects, animals, and birds. Also a mathematical background is mandatory as it deals with mathematical axioms and proofs for the formation of the solution. The problems that are found in the nature require solutions that need lot of computation time and space. The natural environments containing the organisms like algae, ant, bee, fish, cat, wolf, and elephant are already equipped with sensational information that helps them to solve the heterogeneous problems around them. The activities of these organisms when imitated by computers can be treated to be one of the major subjects in this field. Biologically inspired computing is a major subset of natural computation.

Artificial bee colony algorithm is a combination of stochastic genetic/evolutionary algorithm and bee colony optimization technique that helps in solving multi-objective optimization problems. Another type of algorithm is called

E. P. Ephzibah · S. Sree Dharinya (✉)
School of Information Technology and Engineering, Vellore Institute of Technology, Vellore, Tamilnadu, India
e-mail: ssreedharinya@vit.ac.in

L. Remya
School of Mechanical Engineering, VIT, Vellore, India

© Springer Nature Switzerland AG 2020

G. R. Kanagachidambaresan et al. (eds.), *Internet of Things for Industry 4.0*, EAI/Springer Innovations in Communication and Computing,
https://doi.org/10.1007/978-3-030-32530-5_4

swarm intelligence that takes into account the artificial intelligence and the behavioral aspects of cat, whale, chicken, and other animals for solving the optimization problems [1].

Why Do Birds Fly, Bees Swarm, and Fish School?

They do this because the groups are smarter when thinking together than the individual would be on their own. Scientists call this swarm intelligence and this combines the power of many minds to one. Researchers at unanimous AI have developed an innovative technology to allow an online grid to form artificial swarm intelligence. They quickly and actively enter questions, reach decisions, make predictions, generate forecasts, and optimize opinions. A combination of AI algorithms in real-time human input has the knowledge, wisdom, insight, and intuitions and amplifies the collective intelligence empowering them to be smarter together.

A very smart intelligent optimization technique that is new and effective in comparison to the existing swarm intelligent algorithms is called grey wolf optimization algorithm. It is also a swarm intelligent-based algorithm for optimization that mimics the leadership quality of a wolf for group hunting. It is a hierarchy-based technique that is followed strictly to obtain or reach the global maxima [2]. The elephant search algorithm is an algorithm that was built with the idea of searching an elephant, for example, in the case of selecting best gene expression from the microarray data available in large volume [3]. A stochastic feature selection method for reduction of features is also possible with the help of firefly search algorithm. Thus exploring the different AI-based optimization techniques that are biologically inspired for solving problems using machine learning is the objective of the chapter.

To get a deeper insight on the study area, it is mandatory to understand the fundamentals of biological computing or bio-inspired computing.

Neural Networks

Neural networks are data processing algorithms which are adaptive and nonlinear in nature [4]. The network combines the different processing units connected in a system with diverse layers. The characteristics of the networks include its self-adaptability, self-organizing, and capacity to learn from the inputs or response. The neural networks work like neural system in human beings or any higher-level organisms as the way they take input from surroundings. Like neurons in the organisms, neural network replicates to provide output by processing the input. The output from the network is again given back to the network for improvement in the current system if required.

Though there are many approaches to implement neural networks, a widely used and simple one is perception. The perception networks were developed around the 1960s by Frank Rosenblatt. Perception takes several binary inputs and produces a single binary output. The number of input varies from one to any numbers.

For example, a simple perception network will be as shown below:

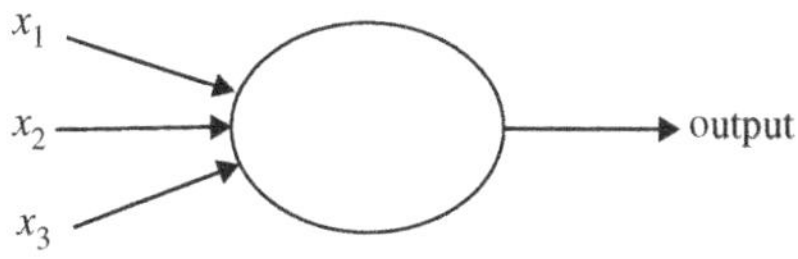

The output of the perception network can be calculated as

$$\text{Output} = \begin{cases} 0 \text{ if} \sum_j w_j x_j \leq \text{ threshold} \\ 1 \text{ if} \sum_j w_j x_j > \text{ threshold} \end{cases}$$

where w_1, w_2, ... are the weights which are real numbers which express importance of respective input to the rate of output. The output label is evaluated based on the value of the $\sum_j(w_j x_j)$. If the value is less than or equal to threshold, the output will be 0, and if it is greater than threshold, output will be 1. Threshold is also a real number which is a parameter of neuron.

We can illustrate it with a real-time application. If the output is buying behavior of a particular product among customers, we can consider input variables as:

- Brand loyalty to the particular product
- Promotion policies that product has currently
- Customer need (whether the customer required the product or not)

Suppose the customer is highly loyal to the brand, $x_1 = 1$. Also there are abundant promotion policies such as free coupons, get one buy one, etc., the second variable $x_2 = 1$. The third is if the customer is in need to buy the product, the third variable $x_3 = 1$. And when it comes to the weight, if we choose weights as $w_1 = 2$, $w_2 = 3$, and $w_3 = 5$. The larger-value w_3 indicates that customer need is important rather than brand loyalty and promotion policies. Finally, suppose we choose a threshold of 5 for the perception neutron. With these choices, the perception will be providing the desired output on the buying behavior of the individual, whether he will buy or not.

The characteristic of perception network is that it can be considered for both linear and nonlinear systems. It can also deal with multiple inputs and multiple outputs. It also includes multiple layers.

Artificial Life

Artificial life or Alife means life made by human being not by the Mother Nature. It consists of the study of the systems similar to the natural system, evolution of such akin systems with the help of computer simulation or algorithms, robotics, or biotechnology [5]. Artificial life acts as a powerful tool for understanding how the ecological communities are evolved [6].

Artificial life researches are mainly divided into three based on its domain:

- Soft ALife
- Hard Alife
- Wet ALife

Soft Alife

It refers to the computer or mathematical simulation. It is a self-organized system and linked to many subdomains. One of the examples for this type is Reynolds' boids model [7]. In it, self-propelled agents move according to kinetic rules.

Hard Alife

It refers to physical robots. Robots are mechanical or artificial agents, guided by the computer program or circuits. Such system has the advantage of the physical reality. Physical experiments are conducted nowadays in many areas including behavior observed in living system, collective navigation, self-assembly, etc.

Wet Alife

It refers *to* chemical or biological research. It consists of the synthesis of life like behavior using the principle of self-organization. A recent example is liquid robot, i.e., dynamic behaviors of microscopically visible chemical droplets.

Cellular Automata

Cellular automata is a field of study that was developed by von Neumann and Ulam in 1960. They are discrete models for complex natural systems that exist in all the areas of life. It is the mathematical representation of physical systems in which space and time are distinct. Also in type of systems, physical measures take on finite set of discrete values [8]. Cellular automata consist of arrays and cells. An array is a regular uniform lattice infinite in nature, whereas cell is the discrete variable at each site. These are also known as cellular spaces, cellular structures, etc.

Learning Classifier Systems

These are the framework that uses genetic algorithms to study learning in rule-based systems. These are multifaceted learning algorithm. It has evolved as the pioneer in evolutionary biology and artificial intelligence [9]. LCS receive feedback from the environment with which it interacts and learns from that (Fig. 4.1).

Learning classifier systems mainly consist of four components:

- Classifier
- Performance component
- Reinforcement component
- Discovery component

Classifier

These are the finite population of condition-action rules. It is the repository of the knowledge of the system.

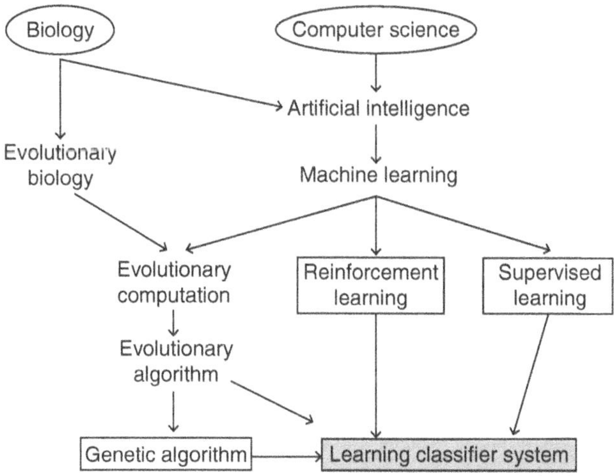

Fig. 4.1 Foundations of LCS community. (Source: Urbanowicz and Moore [10])

Performance Component

It directs the communication with the system and the environment.

Reinforcement Component

It is also known as credit assignment component. It transfers the credit or reward received from the environment to the respective classifier.

Discovery Component

It finds out better rules, and it tries to improve the existing system through genetic algorithm (Fig. 4.2).

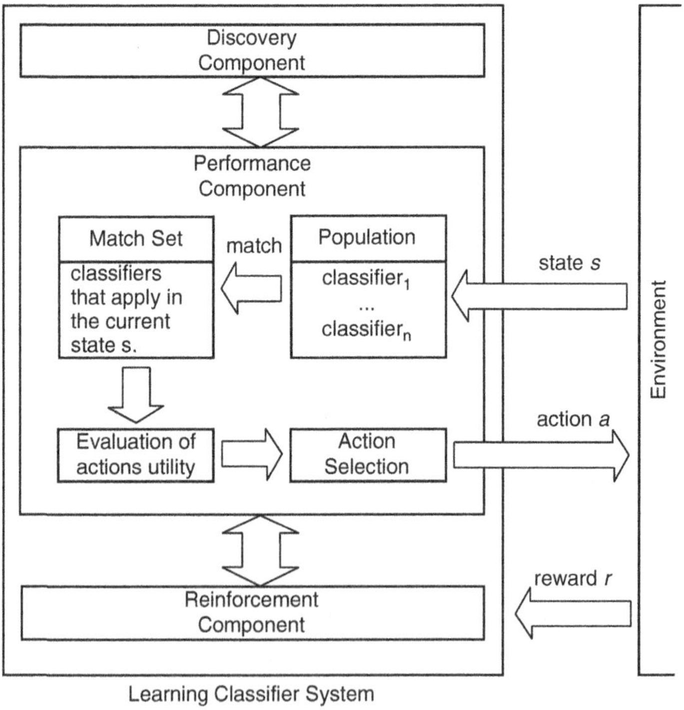

Fig. 4.2 LCS and environment. (Source: Holmes et al. [11])

The main factors that lead to the popularity and acceptability of LCS are its adaptability, generalization, and scalability. While considering adaptability, learning classifier systems are adaptive systems which can handle online learning even in drastically changing environment. It is highly useful in both predicting and describing the drastic changes such as epidemics, etc.

Representation of what it learned and application of learning even if it is unseen situations are the feature of a properly generalized system. In learning classifier system, the evolution of general rules gives way to generalize things especially situations like environmental issues. Scalability is the rapid growth of the system even though it becomes more complex in nature. The methodologies in LCS such as train autonomous robots, theoretical approach for learning complexity, etc. help to increase the scalability of the system.

Models of LCS include Wilson's XCS [12], Stolzmann's Anticipatory Classifier System (ACS) [11], Holmes' EpiCS [13], etc. Its applications are mainly in the areas of autonomous robotics, knowledge discovery, and computational economics.

Membrane Computing

Membrane computing is a group of computation models which are derived from taking inputs and information from the natural aspects such as living cell functioning, cell organization in tissues, organ structures, etc. [14]. Membrane computing was introduced by Păun G in 1998. Several classes of models were defined and were called as P systems. The components of a P system are membrane structure and delimiting compartments. The main issues a P system may face are computing power and computational efficiency.

There are three types of P systems:

- Cell-like P systems
- Tissue-like P systems
- Neural-like P systems

Cell-Like P Systems

It represents the eukaryotic cell. It includes membrane structure and delimiting compartments. The objects are represented by symbols from a given alphabet. The most common type rules are multiset rewriting rules.

Tissue-Like P Systems

Here a common environment consists of many one-celled membranes. It contains huge number of objects, even the environment also. The environment acts as a medium for communication of all cells. Even among them too, few can communicate each other due to the presence of channel. If the cells consist of less rules, it's called P colony.

Neural-Like P Systems

It is of two types. One is akin to tissue-like P systems and cells placed in nodes and contains lot of objects with a state controls evolution. The other is spiking neural P systems which use only one type of object, and information works with distance between the spikes.

The P systems are easily understandable. Its scalability and programmability are welcoming features. Its applications not only restricted to biological process, but to computer graphics, cryptography, optimization, etc.

Sensor Networks

Sensor networks are a group of sensor nodes which are widely dispersed either within the phenomenon or very close to it. The position of such need not be programmed before [15]. Such arrangement helps for deployment in inaccessible areas such as terrains, disaster-prone areas, etc. Self-organizing capability is the main feature to be possessed by the protocol or algorithm. It also has cooperative efforts by the sensor nodes mechanism.

The sensor nodes are scattered in a sensor field. These nodes can collect data and route data back (Fig. 4.3).

The factors influencing the design of the sensor networks are fault tolerance, scalability, production costs, operating environment, sensor network topology, and hardware constraints.

Fault Tolerance

Fault tolerance is the capacity of the sensor network to sustain without any failures in sensor nodes. The reliability $Rk(t)$ or fault tolerance within the time interval $(0, x)$ is

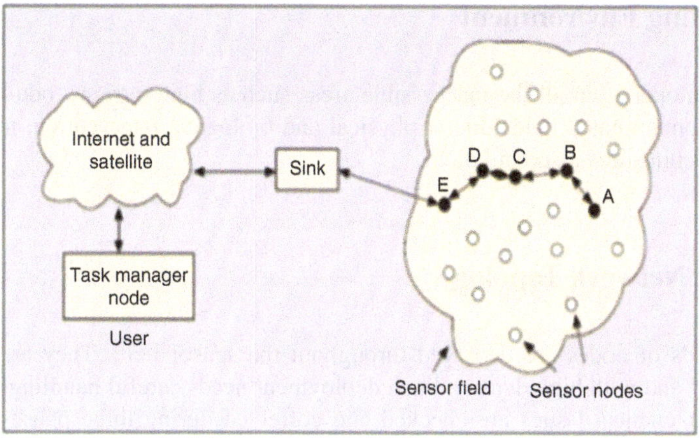

Fig. 4.3 Sensor network. (Source: Akyildiz et al. [15])

$$Rk(x) = e - \lambda kx,$$

where λk is the failure rate of sensor node k and x is the time period.

Scalability

The sensor nodes are huge in number used for deployment. The new schemes must ensure that they are able to work with these high numbers of nodes and can utilize the density properly. The density μ can be calculated as

$$\mu(R) = (N \cdot \pi R2) / A,$$

where N is the number of scattered sensor nodes in region A and R is the radio transmission range. Basically, $\mu(R)$ gives the number of nodes within the transmission radius of each node in region A.

Production Costs

Since the sensor nodes deal with large number of sensor nodes, cost of each sensor will be an important factor which justifies the overall price of the product. If the overall cost of sensor networks is higher than the traditional network, it won't be cost justified. And hence the cost of each sensor node needs to be kept low.

Operating Environment

They can operate in all the inaccessible areas such as high terrains, bottom of the ocean, contaminated fields due to physical and biological intervention, interior of large machine, and also battlefield.

Sensor Network Topology

Thousands of nodes are deployed throughout the sensor field. They are densely deployed and with high density. Such deployment needs careful handling of topology maintenance. Issues are checked and corrected during three phases such as redeployment and deployment phase, post-deployment phase, and redeployment of additional node phase.

Hardware Constraints

Sensor node has four components – a sensing unit, a processing unit, a transceiver unit, and a power unit. It has additional application-dependent components such as location finding system, power generator, and mobilizer. Sensor unit consists of sensors and analog to digital convertors (Fig. 4.4).

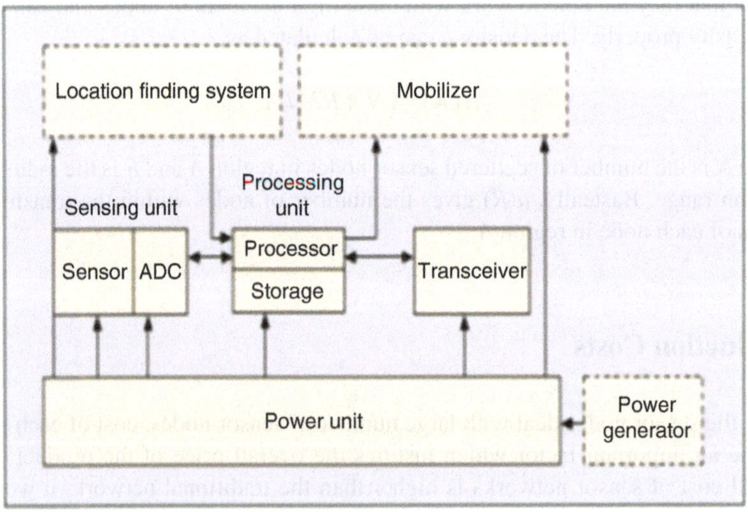

Fig. 4.4 Sensor node components. (Source: Akyildiz et al. [15])

All these subunits need to fit in very small-sized module smaller than even cubic centimeters. There are constraints apart from size. The nodes must consume extremely low power, operate in very high densities and low production cost, be autonomous, operate in inaccessible areas, and be adaptive to environment.

Sensor network has many applications. They are part of military command, control, communications, computing, intelligence, surveillance, reconnaissance, and targeting (C4ISRT) systems. The rapid deployment, self-organization, and fault tolerance characteristics of sensor networks make them a very promising sensing technique for military C4ISRT. It has environmental applications such as forest fire detection, bio-complexity mapping of the environment, flood detection, and agricultural application. While coming to health sector, it can be used for telemonitoring of human physiological data, monitoring of people inside hospital, and drug administration in hospitals. In home applications, its service extends to smart home and home automation. Other applications include vehicle theft detection, vehicle tracking and detection and interactive museums, etc.

Biodegradability Prediction

Biodegradability prediction tries to predict biodegradability of anthropogenic materials in the environment. As it became very essential to predict the impact of the chemicals in the environment, different computational models are taken into consideration for the same [16]. Modelling techniques include linear and nonlinear regression, expert system, neural networks, and artificial intelligence.

Regression Models

It consists of liner, nonlinear, and multilinear correlation of biodegradation rates with parameters such as molecular weight, structure, solubility, etc.

Expert System or Survey Models

This model stakes expert opinion in the respective area by conducting surveys, and based on that models predict degradation.

Machine-Based Learning

Including neural networks, these models process the chemical structure of the products based on it.

Biologically inspired computing which is commonly referred to as *bio-inspired computing* is a close relation to the field of artificial intelligence which has a link with evolutionary algorithms. Evolutionary algorithms form the basis for a very impressive and beneficial evolutionary theory with a badge called *survival of the fittest*. That means answers to a composite problem can be obtained by using the computer as a means to get inexpensive ways to deal with complex problems. Particularly they are used for search and optimization problems.

Swarm intelligence is a special phenomenon that groups all the different types of algorithms that take into consideration the term called *communication* as its key mechanism. The living creatures together put their efforts and cooperate and successfully solve problems because independently they will find it tough to elucidate. But if these organisms join their efforts, they can solve these types of problems. Swarm intelligence algorithms are very good and efficient in solving optimization problems. Moreover, they are useful in classification, clustering, robotics, and other types of biologically inspired computing applications.

Some examples of swarm intelligence algorithms are particle swarm optimization which is inspiring some behavior of bird flocks, ant system, the bacterial foraging optimization algorithm which is based on the foraging behavior of the *E. coli* bacterium, and bee colony optimization.

Dynamic constrained optimization problems are interesting to solve as the solution to be obtained is dynamic or movable in nature. So the algorithm needs elements to track the moving optimum. The problems become quiet complex in some steps of the search. Robotics is a field started in industrial settings where very precise and fast motion or dynamic characteristic of the machines is needed. Those technologies are not suited for interacting with people.

Bio-robotics is a field of artificial intelligence where thinking is involved in the combination of both body and the brain and the group. There exist a lot of examples in biology where we can see how that can be achieved and how we can leverage the same kind of ideas. It is a new discipline connecting living creatures and robots. By looking at the nature, how nature does things. Looking at on the design principles of nature and trying to build robots that are either implemented through those principles or act like those principles. Looking at the materials that exist in nature and the algorithm that nature uses, the behavior and the activities of the animals can be used to learn from them and use it applying those techniques to robots.

Almost all the locomotive robots are inspired by these soft animals that live. Compared to more traditional robots, something that are used for welding the doors of cars, robots are made to enhance safer technology. Because they are soft and small, because of their size, and because of their compliance body to adapt to natural environments, they can perform tasks that are difficult and impossible for other classes of robots that are conventional in nature. Many of the people in this field are inspired by the incredible ability of the biological systems to cooperate. From the nature with swarm of insects like termites that build mount or ant colony that forge over large regions, their quick and wise decision-making skills inspire to build the robot systems that could do all tasks efficiently. Could I make robots that build, and could I make robots that could go through coral reefs and monitor the house of a system? Consider the traffic on the highways and roads, and self-driving cars, a lane full of self-driving cars, there are a lot of robots on the road.

An algorithmic understanding can be built and large number of skilled laborers can work together in a decentralized way. And they can be slightly faulty, and they can have low capabilities as opposed to perfect robots that navigate through complex environment. That's really amazing manufacturing technology because mass manufacture of simple things is possible and these simple things will cooperate and create something of much higher complexity. What we are aiming at in robotics is much really where humans and robots can coexist, can coordinate, and can distribute tasks to the joint benefit of all. As you know there are many big differences between driving mechanism of current robots and human body. Work on building soft robots is being carried out to enhance and augment healthy human individuals. These robots will give people with disabilities the super human strength. There are a lot of people with some kind of disability that need human strength but just need a small system. New robotic components are manufactured that are using soft material able to apply to person with mask and associated with digital devices.

Many animals have the ability to move to newer and dynamic spaces that are unpredictable in ways that engineered systems cannot achieve even with human operator. But if we did have these systems that can achieve these capabilities, we will have much better ability to address needs such as rescue, emergency response, and collecting information from environmental monitoring. Using fundamental biological principles, some of these challenges can be addressed in order to improve engineering design and operation.

The fish robots are one of serious applications. Consider the task of environmental monitoring on a global scale to collect information to make better models of what's going on in the ocean. Given information on temperature, humidity, chemical content in the ocean, and species growth and decay and to come up with better ways of knowing what's going to happen in terms of seismic activity, species tracking, weather currents, and transportation are safe. The ways of data being collected urgently involve sending the research crews with number of people, and they go to specific locations. Collecting the data using autonomous vehicles is the need of the hour. Fish robots were developed to operate in dynamic spaces, not very well known and changing very quickly to be sent for a period of months. Migrating to the remote location to fix parts or physically collect the data that come out with them has become troublesome. This can be done with energy-efficient sensors that have lot of power to run, with high-power camera to show what is going on in a large space. The engineering characteristic to achieve with similar capabilities in biological systems could be a fairly large characterization here. The robots should be fast and agile. One of the fastest swimmers in nature is tuna fish. The key thing about them is that they have a fairly narrow body, the first part is fairly long and rigid, and then there is a fairly rigid tail fin at the back. So a model of a fish with a rigid body in water is taken as the model, and the characteristics are noted for lift and drag operations and activities. This can give the information about how to make the vehicles go quickly in a large remote and unknown space, whether it is under the ocean, inside a dense forest, or in the space.

Role of ant colony optimization technique in decision-making for robots:
Decision-making using ant colony optimization is an emerging trend in the field of robotics that tries to analyze and enhance the path distance between any two positions. The robots mimic the behavior of ants for motion. The ants move in search of their food from their resting place to the spot where the food source is available. Randomly, ants take the first move. When they walk on the floor, they leave pheromone trail, a chemical substance that is capable of getting evaporated in due course of time. Subsequently, when other ants get into that space/location, they find these pheromone trails, and based on the intensity of these trails, they try to learn the distance to the food particle. The chemical substance reduces the intensity when there is less number of ants or no ants in the path. The intensity is more and vigorous if the path is less and there are more ants going and coming back from the source to the destination in that path. Obviously the ants try to follow the way where the intensity of the trail is more, thereby educating the ants in this phase and helping them to take decision. This gives the major idea or information to the robots in the case of robotic movements as well. Robots, in which many in number work in the same workspace, in parallelism and simultaneous manner, are made to perform their tasks without colliding with each other. Robots in the workspace are expected to move communicating each other like ants in an efficient and effective manner. Communication among robots plays a very important role in building the system. They learn from each other to achieve the target. When the robots get the information about the path travelled by the other robot, they receive information in terms of time, energy, etc. Hence decision-making becomes effective and the shortest route is also identified. The data received influences the other robots to follow the path travelled adequately by their predecessors who have created a pattern-like structure for efficient building of the system. The ant colony optimization (ACO) method firstly has the arbitrary path in the starting of the method, and they get converged into an operative one with minimum energy so the robots will follow it most effectively using ACO, an algorithm that takes advantage of the functioning of the ant colony.

Genetic bee colony optimization for charging unmanned cars:
Bees are very smart and probably the smartest insects, and they are the source of solution for lot of real-time problems. Bees are extraordinary explorers and navigators. They orient themselves in a radius of 6 kilometers around their house. They remember numerous resource locations like food, water supply, and their hives, and they commute between those locations in astounding accuracy and that with a brain that is of a very small size of a peanut. But the individual feet of the bees are the only one of the many facets of bee intelligence. It is together how they solve complex tasks; they act collectively as a network of tiny brains looped together to vari-

ous forms of communication. They share information and also energy. The process of sharing turns one centralized static energy store into many thousand small mobile energy store. Electric cars need careful planning and strict discipline. Substantial infrastructure changes have to be made for charging them to meet the energy hunger for tomorrow's cars.

How are we going to make our electric cars convenient, cost-efficient, and environmentally less harming?
The obvious answer is that we charge them like bees. We need three components. Firstly, cars need to ask another car for energy, secondly cars need to be charged, and thirdly self-driving cars have all the sophisticated sensors in them with computing equipment. There will not be a necessity to stop the car to charge between them. We can charge our neighbor's car while driving, and this is how it's going to work in the near future. Cars can communicate between each other and do a quick and safe energy transfer between moving cars. There could be many potential donors thousands of charging possible along the road. Cars need not be charged to full, or even some charge can be added or received as much as consumed. As long as the target is reached and as long as the other one arrives at his or her target location, everyone would be happy to help because the swarm has become energy safety net. The swarm is like a virtual power cable extending from the wall with the plug through many cars to the ones that actually need the charge. Cars are sort of mobile power outlet, and if traffic jams are common, they also can now be reduced to some extent. This could even be the business concept. The car manufacturers are informed to put this technology into the cars. Lot of money can also be generated by this kind of technology and people can be helped. By the year 2022, 50% of the car equipped with this technology is expected, and if just charge 10% of the energy that is being transferred as a feed for this service, and an average number of ten energy snacks per month would yield the market of seven million euros in Germany alone, and now think of Europe, china, or the USA. This would help the customers who would like to operate their cars all the time without any stops. Those people that charge their cars regularly provide service to one another. Think of self-driving, taxi companies, or delivery companies, they would be able to operate their cars throughout the day and night; this technology would definitely be more beneficial.

References

1. Darwish, A. (2018). Bio-inspired computing: Algorithms review, deep analysis, and the scope of applications. *Future Computing and Informatics Journal, 3*(2), 231–246.
2. Sharma, P., ShirshSundaram, M. S., Sharma, A., & Gupta, D. (2019). Diagnosis of Parkinson's disease using modified grey wolf optimization. *Cognitive Systems Research, 54*, 100–115.
3. Panda, M. (2017). Elephant search optimization combined with deep neural network for micro-array data analysis. *Journal of King Saud University - Computer and Information Sciences*, In press.

4. Graves, A., Wayne, G., Reynolds, M., Harley, T., Danihelka, I., Grabska-Barwińska, A., et al. (2016). Hybrid computing using a neural network with dynamic external memory. *Nature, 538*(7626), 471.
5. Langton, C. G. (Ed.). (1997). *Artificial life: An overview*. Mit Press United States of America.
6. Maes, P. (1995). Artificial life meets entertainment: Lifelike autonomous agents. *Communications of the ACM, 38*(11), 108–114.
7. Reynolds, C. W. (1987). Flocks, herds and schools: A distributed behavioral model. *ACM Siggraph Computer Graphics, 21*(4), 25–34.
8. Codd, E. F. (2014). *Cellular automata*. Academic Press Academic Press, Inc. Orlando, FL, USA.
9. Lanzi, P. L. (2000). *Learning classifier systems: from foundations to applications (No. 1813)*. Springer Science & Business Media, Berlin Heidelberg.
10. Urbanowicz, R. J., & Moore, J. H. (2000). Learning classifier systems: A complete introduction, review, and roadmap. *Journal of Artificial Evolution and Applications, 2009*, 209–238, Berlin Heidelberg.
11. Holmes, J. H., Lanzi, P. L., Stolzmann, W., & Wilson, S. W. (2002). Learning classifier systems: New models, successful applications. *Information Processing Letters, 82*(1), 23–30.
12. Wilson, S. W. (1987). Classifier systems and the animal problem. *Machine Learning, 2*(3), 199–228.
13. Holmes, J. H. (1996). *Evolution-assisted discovery of sentinel features in epidemiologic surveillance*. Drexel University.
14. Păun, G., & Rozenberg, G. (2002). A guide to membrane computing. *Theoretical Computer Science, 287*(1), 73–100.
15. Akyildiz, I. F., Su, W., Sankarasubramaniam, Y., & Cayirci, E. (2002). A survey on sensor networks. *IEEE Communications Magazine, 40*(8), 102–114.
16. Peijnenburg, W. J., & Damborský, J. (Eds.). (1996). *Biodegradability prediction* (Vol. 23). Kluwer Academic Publishers, Petra Sprado.

Chapter 5
Internet of Things and Additive Manufacturing: Toward Intelligent Production Systems in Industry 4.0

A. Suresh, R. Udendhran, and G. Yamini

Introduction

The Internet of things, Industry 4.0, and smart sensors can be combined to make a heady cocktail that suggests an exciting future with effective and improved performance. Security is very essential in every field form in industrial sector. Through an intelligent system, we could able to detect major challenges existing in Industry 4.0 and address it.

A sensor is an electronic device that senses the physical quantity and converts it into electrical quantity. They can be used to sense pressure, temperature, humidity, etc. For the purpose of interpretation and processing, the sensor data could be taken and then displayed in the computer for the human reference purposes. Nowadays, sensors are widely used for the protection of facility control and industries as shown in Fig. 5.1.

Traditional manufacturing process works on the principle in producing certain number of products by holding the reserve in case of unexpected damage or shortages at certain set of periods [1]. There are four main categories involved in the traditional or commercial technology. These are as follows:

Injection Molding: As the name refers, a plastic material is softened, and then it is injected during the fabrication process inside a mold [2]. This process comes under a mechanism, and the other mechanism involves the removal of the material from the mold after which it could be cured and solidified. This is highly used in the plastic goods fabrication process that gives a good finishing surface.

A. Suresh (✉)
Department of Computer Science and Engineering, Nehru Institute of Engineering and Technology, Coimbatore, India

R. Udendhran · G. Yamini
Department of Computer Science and Engineering, Bharathidasan University, Trichy, India

© Springer Nature Switzerland AG 2020

G. R. Kanagachidambaresan et al. (eds.), *Internet of Things for Industry 4.0*, EAI/Springer Innovations in Communication and Computing, https://doi.org/10.1007/978-3-030-32530-5_5

Fig. 5.1 IoT sensors in industry

The disadvantage among these processes is that the cost of the injection mold is very high and it is not sufficient for the low-budget series.

CNC Machining: The material should be machine as a part in any field that involves drilling, turning, or milling [3]. In this process, the machine is fitted with the material particle, and several tools could be used to remove the material until a desired format is obtained. Similar to the previous process, this process could not be used for the small-scale applications due to their cost. But the accuracy of this process is very high with the tolerance rate of 25 μm. But the internal features could be fabricated with this process and are very tedious for the machining process [4]. It is best suited for the engine components.

Plastic Forming: Several techniques such as pressure forming, thermoforming, and vacuum forming are involved in this process. Each type is specific in their functionality, but the process of heating a plastic sheet and draping it around the mold is similar. Then the main plug and the sufficient air pressure could be used to make a shape out of the sheet. The forming process utilizes thermoplastics virtually, and only one side of the plastic could be modulated by this process [5]. Compared to the above two processes, the installation and by-product costs are very low.

Plastic Joining: This process involves in combining the semifinished parts. Welding, adhesive bonding, and fastening are the process involved here [6]. The incorporation of latches along with the snap and hinges that fits into the design part or with the screws and bolts as the external fasteners comes under the process of fastening. Epoxy is said to be an adhesive that could be used in joining the parts together involved under the category of adhesive bonding [7]. Finally, welding is the stage that involves the application of heat under sufficient pressure to combine the material together. The semifinished parts and their property play a vital role in choosing the sufficient technique [8]. However, this process is quite time-consuming, and the cost for the labor charges seems to be very high [9].

Comparison Made Among Traditional and Additive Manufacturing (AM) in Industry 4.0

Based on all the abovementioned process regarding the AM and traditional manufacturing process, a comparison table has been given that could help in choosing the right and adaptable one for the process [10] (Table 5.1).

Based on the application, the best suitable method should be adapted, and identifying the requirements is the most necessary matter of all [11]. These are the following requirements:

Quantity: Large-scale manufacturing process could be adapted by the injection molding and foaming. On the other hand, AM methods [12] are best adapted for the small-scale process. The cost estimation had been made for the GoPro handle that has been carried out with the injection molding (Quickparts) and SLS (Sculpteo) process, and it shows the SLS technique is the most effective valid option at 486 units as shown in Fig. 5.2.

Table 5.1 Comparison of different IoT-enabled processes with the advantages and disadvantages in industry

Process	Advantages	Disadvantages
Selective laser sintering Layers of 0.06–0.15 mm with the fast print speed and rough surface	Very complex and strong Possible snap features and the parts could be stacked in the volume	Rough surface with the grainy finish
Stereolithography Layers of 0.06–0.15 mm with the average print speed and smooth surface	Finished end product will be very smooth	The parts are very weak and less sustainable to heat and sunlight
Binder jetting Layers of 0.089–0.12 mm with the very fast print speed and rough surface	Print speed is very fast with the multicolor printing facility	Surface finish is very rough with the weak parts and quality
Fused deposition modeling Layers of 0.1–0.3 mm with the low print speed and very rough surface	Strength of the whole setup components are very high with the minimization of cost	Poor-quality surface and printing speed is very low
Injection molding Excellent surface finish with the tolerance level of 50 μm	Reasonable turnover rate with the high tolerance	Equipment cost is very high and parts of the equipment are really thin
CNC machining Smooth surface finish with the tolerance level of 25 μm	Materials are very compact with the reasonable turnover rate and high tolerance	Equipment cost is very high and complex design
Plastic forming Smooth surface with the 1 mm tolerance value typically	The price is very nominal for the large body parts	Single-sided control with the equipment components really thin and uses only thermoplastic as their material
Plastic joining Based on the finished product	All the materials could be used in this process	Labor cost is very high and huge time-consuming process

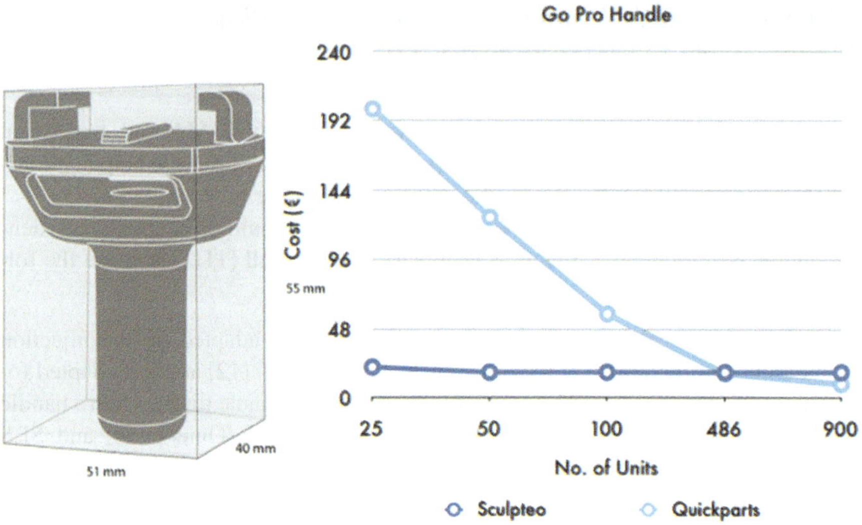

Fig. 5.2 An example of AM in Industry 4.0 with injection molding

Lead time: In case of the traditional manufacturing process, the number of days for the units to be manufactured could be increased due to the arrangement of molds, but this is not the case in 3D printing. In case of the GoPro handle, 25–1000 units could be done at a period of 15 days, whereas 25 units could be done within 3 days, and 1000 units could be completed within 7 days with AM process.

Complexity and shapes: Customizing the complex parts is the essential one that could be done effectively by 3D printer using a professing [13]. This could not be effective when it comes under molding, forming, and machining.

Selection of material: Traditional manufacturing leads to the selection of high-quality material, but when it comes to the 3D printer, the material selection could be scattered or thinly dispersed [14]. In certain cases, SLS uses powder made up on thermoplastic material that is specific to that particular machine, FDM uses molten thermoplastic, etc.

Other considerations: Several applications that use the 3D printing technology with their tensile strength [15].

The Role of Additive and IoT in Industry 4.0

The creation of digital representation of the object is extremely important with its geometrical attributes, texture, and color. This role is successively played with the help of additive manufacturing (AM) and subtractive manufacturing (SM) [16]. The traditional way of trimming of materials with the help of high-end tools is known as SM [1]. AM process effectively used the software platform that is commonly found in visual aids, tooling components, functional models, direct manufacturing, etc. [2]. The process of cutting away the material from the certain block of material to bring that block to a required shape is termed as the subtractive manufacturing process [3]. This traditional process could be done manually, but accurately this could be done through the CNC machine [4]. As per the requirement of the designers, the CNC machine could perform the cutting task, which is concentrated on the three axes (say x, y, and z plane) of the material. In doing so there could be a less chance for the designers to flip the material. There is a major advantage of adapting subtractive manufacturing since this process has the ability to extremely machine a miniature piece of material into a living hinge [5]. This process could not be achieved with the 3D printing. Certain parts could be created with the help of the additive manufacturing that needs living hinge component, while special components could be created with the help of a living hinge [17]. The most commonly used devices in the process of Industry 4.0 manufacturing have been discussed below:

Roland 3D CNC Milling Machine MDX-40A15

The Roland 3D CNC Milling Machine had been especially designed for the process of rapid prototyping as a benchtop CNC machine [7]. This particular 4-axis machine could handle several types of materials and is said to be G-code compatible. A smooth finish could be provided as an end product. The cost of the material is extremely low [8]. This milling machine is used moderately in industries. Digital caliper finds its place in the majority of the measurement purposes. This could promote maximum accuracy with the precise distance of measuring small objects. These calipers permit for "zeroing" at any point along the slide, which could permit differential measurements [9]. The Roland 3D CNC Milling Machine MDX-40A15 along with the digital calipers has been shown in Fig. 5.3.

(a) (b)

Fig. 5.3 (**a**) Roland 3D CNC Milling Machine MDX-40A15. (**b**) Digital caliper

Additive Manufacturing Process in Industry 4.0

Additive manufacturing popularly said as rapid prototyping or 3D printing has gained its attention recently that uses the principle of addition of materials in a layer with the particular processes [18]. This requires an additional material to be fitted which seems to be designed with the help of certain software in a three-dimensional structure. This can be directly implemented in the real-time application purpose, thereby decreasing its time and the complication. Instead of removing the materials, the addition of materials serves the system as a better head conductor and makes them lighter [11]. This could remove certain constraints that are created by the conventional manufacturing by adding the materials layer by layer [12]. At first, the materials are added in a thin powdered layer to the building platform, and with the help of the laser, the powder diffuses in certain points by a computer-generated component design data. Then another layer is placed that fuses the material to form a predefined structure. The additive manufacturing technology can be used in wide variety of fields starting from the medical and marine application to the electronics and consumer accessories [19] as shown in (Fig. 5.4).

Several processes are involved in a metal additive manufacturing and they are as follows [14].

Material Deposition Process

This method transports the material from the powder jet or the filament into the surface. The depth of the layer could vary in a certain micrometers range. The

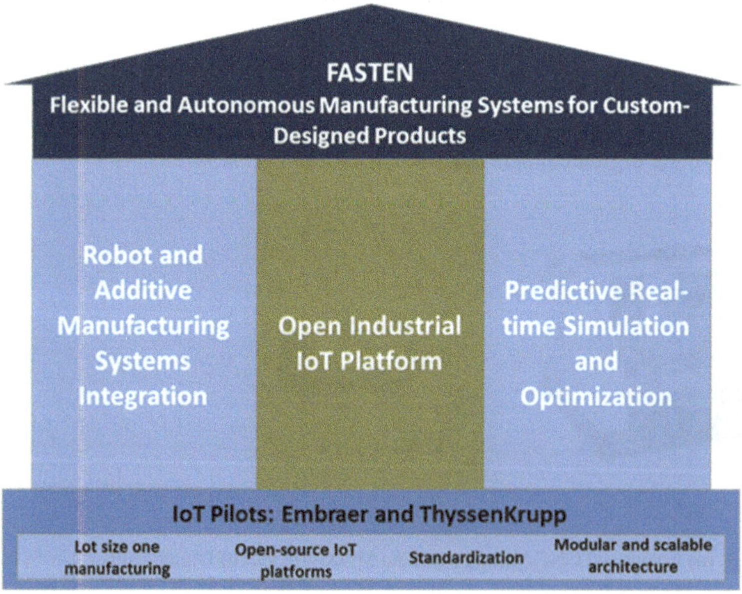

Fig. 5.4 Block diagram of the stages involved in the metal additive manufacturing

important factor monitored over here is the flow rates of the material either a powder or a polymer [15]. Depending on the powder size, the flow rate should be estimated. The material is supplied in 2 axes although it's a 3D axis. With the help of this process, we could eliminate the usage of any support material that does not cause a viable geometry [16]. Material deposition could be done with the help of cold spray, extrusion process, and blown powder process.

Power Bed Process

(a) *Laser Beam Melting*

Additive manufacturing involves several ways in the deposition of the layers. Selective laser melting (SLM) and selective laser sintering (SLS) process melts and softens the layer settled in the material. Certain industries use the metal powder or the polymer in the production of the layers. In the process of laser beam, the high-intensity laser is used to melt the metal. Melting process involves the material of powdered form [20]. The powder is spread equally with the help of re-coater arm, and the laser power of about 50–2000 W has been used to solidify the process. When this is done, then another layer of powder is coated, thereby lowering the depth of the material. This particular process is carried out in an inert temperature of argon or helium atmosphere since the powder can easily oxidize. When the depth of the layer is small, then the manufacturing part is well defined, although the time of manufacturing increases. A heat posttreatment is necessarily carried out to remove the deformation of the materials or polishing the surface [21]. This should be done in a controlled environment in order to increase its precision rate. Similarly, the material used in the laser polymerization is the photosensitive resin. This process is cured with the help of the UV radiation from the low intense laser source.

(b) *Electron Beam Melting*

This process is more productive when compared to the laser in its features. This process uses an electron beam of maximum power of about 3 kW. This is also carried out with the certain temperature in an inert gas environment. The powder used should be preheated that neglects the thermal gradient. Although the temperature used here is very high, the cooling procedure is also carried out to avoid the internal stress of the material. The powder could move at a faster rate that works well with the huge depth of the layers. But due to its broad beam width, this method is not precise, and it is not sufficient when compared to the laser beam method [22].

Liquid Bed Process

This involves the process of stereolithography and thermosetting resin. The principle of photo-polymerization is involved in the stereolithography procedure. The lower-power helium-cadmium UV radiation has been used in this process [23]. This UV radiation coagulates to form a thin layer in the surface of the material. A SLA machine is built in such a way that the whole process is engrossed in a liquid resin with the laser source and the required hardware and software platform is also built. With the CAD model, the layer is being scanned, and once this process is done, the

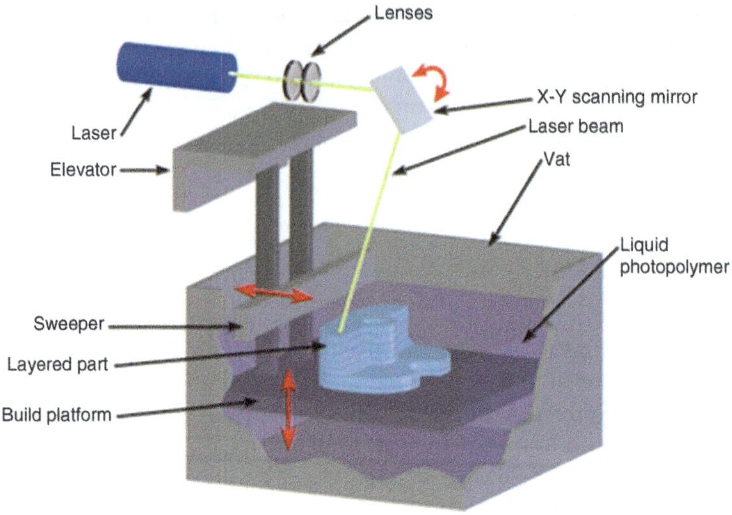

Fig. 5.5 Parts involved in IoT-based stereolithography

inner portion is engraved and undated below a layer [24]. A blade is used to remove the extra fitted layer to ensure the flatness of the surface. After this process the next layer is deposited, and again the procedure is carried out like the previous one. With this method we can able to produce a highly accurate polymer parts [25]. The process of stereolithography is shown in Fig. 5.5.

IoT-Based Additive Manufacturing Versus Traditional Manufacturing in 3D Printing

For comparing both the IoT-based AM and the traditional manufacturing process, we have defined a part that has been printed by both the methods. Figure 5.6 shows the block printed by additive manufacturing process. The block has a well-defined feature and finish. The CNC milling machine encompasses a much diverse process than the 3D printer [26]. Sufficient supervision should be provided for carrying out the process at CNC machine though similar files are being used for making the part. This is because the materials are removed from the block during the process. This part is sufficiently made with the help of wax. Previous studies have stated that the CNC machine could provide best finish or end product when compared to the 3D printer [27]. Figure 5.7 shows the block created by the CNC machine. This particular block has certain problem when this case is closely compared. The inner crevice of the M has not properly curved due to the miniature dimension and the size of the tool used for cutting. We could change the drill-bit of the machine by replacing it with the smaller one. But this could increase the complexity of the machine, thus increasing the processing time [28].

Fig. 5.6 Block created by the 3D printing

Fig. 5.7 Block created by the CNC milling machine

After the manufacturing process, it is necessary to compare the measurements with the certain parameters [29]. There is a necessary usage of the scale factor that compares with the original model, and this could be due to the occurrence of size restrictions. The end product of the CNC machining material could be slight thicker in their dimensions, whereas with the 3D printing, the end product could shrink from their original size due to their plastic material shrinking after the process of printing and cooling. Several parameters that are sufficiently measured for both the SM and AM process are as follows:

- Setup – Number of steps and setup time for each machine
- Ease of use – how easy is it to use the machine
- Percentage of wastage of the material (%) – wasted material due to the process of printing and cutting
- Machining speed – Manufacturing the parts rapidly at certain time period
- Accuracy – Comparison of accuracy made with the model
- Surface finish – smoothness of the part surface

Discussing about these parameters, it is said that the surface finish of AM is comparatively low when compared to the CNC method [30]. In all the other parameters, additive manufacturing leads its place that could be modeled with the help of CAD software platform. In fact, several trials could be possible only by the AM process. Surface quality is the most important feature to be considered in the additive manufacturing. Depending on the particle size, the surface should be used in manufacturing. When the particle size is huge of above 30 μm, then the surface produced by the selective laser melting will be coarse. However, less particle size of below 20 μm could produce a smooth and a polished surface. There is a drawback in using the fine-sized particle as it could hinder the flow rate of the particle in the layer, thereby increasing the cost. The surface quality is the most important concern in the quality estimation. Certain process involves the fine powder layer, for instance, binder jetting process (BJG). The tolerance level should also be taken into consideration due to its heat treatment in its fabrication process. If this fails, then there could be considerable dimensional changes in the end material. To avoid this fault, additional necessary acceptable materials could be used which could be removed at the later. Besides, the process of SLM and EBM is more accurate with a reduced amount of tolerance rate. Hence, with the availability of the resource, product production time, number of products produced, etc., we must select the method based on these requirements. The process selection is done with the help of the type of the material we use during fabrication. When the ceramic type of material is used, then BJG is the most preferred process that could be used. When the fabricated material is of metals or composite type, then we can adopt any three processes like SLM, EBM, or BJG. For example, let's consider the magnesium-based materials. Due to its low boiling points, the process of EBM could not be adopted. Due to their high intense beam used in their power bed, which results in the melting of certain materials. Therefore, the best adopted method could be either SLM or BJG. In the fabrication of brittle materials such titanium aluminide, the three processes could also be adopted. But due to the high cooling rate, SLM process could not be adopted since this could lead to the internal stress of the material. The cracks could be eliminated only when the cooling rate is very low. The material limitations and their property are shown in Fig. 5.8.

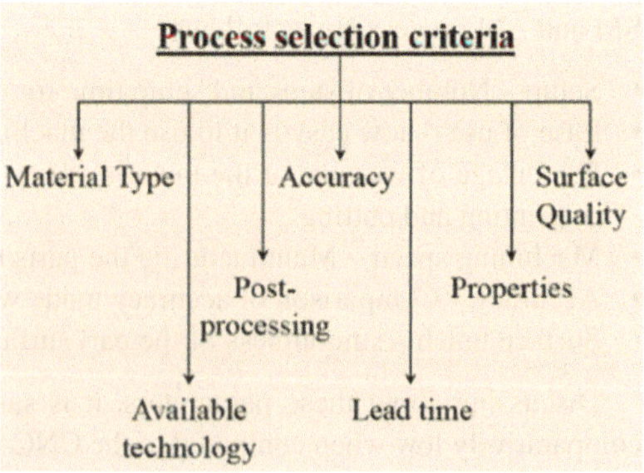

Fig. 5.8 Process selection criteria

Advantage and Faster Deployment of IoT-Based Additive Manufacturing in Industry

In the aspect of the capabilities of manufacturing, a different trade-off between AM and CM exists among the processes in which the additive manufacturing possesses the capability of reducing materials as well as usage of energy, less wastage, innovative acceleration, as well as reliable supply chains [31]. Therefore, it is evident from the above statement that AM plays an important role in deciding the future of manufacturing. Let us discuss the important advantages of the additive manufacturing.

Creativity: By integrating CM processes, the AM leads to innovative frameworks along with its effective geometrics which can be easily developed during the course of the manufacturing process. This type of innovation enhances the performance, environmental-friendly, as well as cost-effective solution in engineering and product application, for instance, AM systems which are programmed to accommodate properties inside a component which enhances the performance by employing conventional processing.

Design products with fewer, more complex parts: The unique ability to create components using few but complex parts is a major advantage of AM. This feature is so unique that it only exists in AM which upgrades the performance at a system level excluding the manufacturability present at the subsystem level. In order to reduce the overhead association concerning the production planning as well as control and documentation, a minimum number of parts in an assembly are considered. By employed minimum number of parts, less labor time needed to manufacture the product and the "footprint" of the assembly line are gained which in return lessen the overall reduction in manufacturing expenses.

Preserve energy by obviating production process: Lower energy consumption is achieved by deploying AM which results in less material usage, encouraging recycle of by-products, and manufacturing lighter products.

Decrease in wastage: Creating objects layer by layer reduces materials needs and less human error in production as well as the costs by 96% unlike the traditional manufacturing working that increase usage of material. Eventually, "cradle-togate" present in the environmental footprints of parts manufacturing through rejection of the tools diminishes as well as material scrap existing in the CM processes.

Faster deployment to Industry 4.0: As soon as the Standard Tessellation Language [STL] file is created by employing 3D digital description (3D scanning or 3D imaging) to the parts, the need for more expensive, in-effective part tooling as well as prototype fabrication is achieved.

Less weight: With less material, complex shapes applied to the specification of the conventional parts with the help of the AM can be achieved.

AM enables rapid response to markets: With the ability to build fresh production options outside of factories, for instance, mobile units that can be located close to the source of local materials, thereby increasing the agility of manufacturing. AM does not include high-cost tools unlike in traditional manufacturing which leads to longer production process and slow production rate when there is need for high volume of production. AM does exist in the manufacturing operation to fabricate the molds, as well as tooling employed in the production.

In order to deploy the tooling faster to market with less cost, tooling is created as well as printed directly to the market and not outsourced. Spare tooling parts can be manufactured on demand, excluding the need for stockpiles as well as complex supply chains. However, one challenge encountered in AM tools in manufacturing metal tooling is the experience of higher capital costs as well as steep learning curves which should be addressed by increasing the rate of consumption by the manufacturers.

Case Study

By employing injection molding, numerous manufacturing cost data for five unique objects, namely, a cockerel, a GoPro handle, and representative models of products, were gathered. The below table represents the five major objects along with their bounding box dimensions: This section of case study presents a vivid understanding of the Sculpteo batch control 3D printing which is more costly unlike the rapid injection molding for industry representative parts.

The details regarding the cost of five parts in 3D printing production can be found in the Sculpteo website. There are several algorithms which can assess the part for 3D print-ability for any errors which lead to failure of 3D printing. Moreover, "Solidity Check" can be organized by the Sculpteo which assures that all behavior of the part possesses minimum feature size threshold of approximately 1 mm. The calculation for cost per unit and its total batch cost can be determined by employing the "Batch Control" option present in the selective laser sintering.

By producing effective production run, the batch control can be automated for more than ten parts which optimizes the unit price by producing effective solutions which include orientation, layer thickness, and finish quality which leads to reliable control and well suited for separate products. By employing the "rapid injection molding" method, the production components can be produced based on the quotes from the manufacture and mainly employed for prototypes as well as short series manufacturing, but to a limit of 10,000 units. But the parts which need hollow interior do not need this technique. However, for parts with hollow interior, "gas-assisted injection molding" can be employed which is considered as secondary option for manufacturing.

The charts represent the comparison among the injection-molded technique and the 3D printing techniques. The number of manufactured units is being mentioned by the intersection point in the chart as shown in Fig. 5.9.

The data gathered presents the 3D printing which offers insight into the economical needs to manufacture groups of approximately 500 units more cost-effectively unlike the rapid injection molding methods. The injection-molded parts

Fig. 5.9 Comparison chart for various manufacturing parts

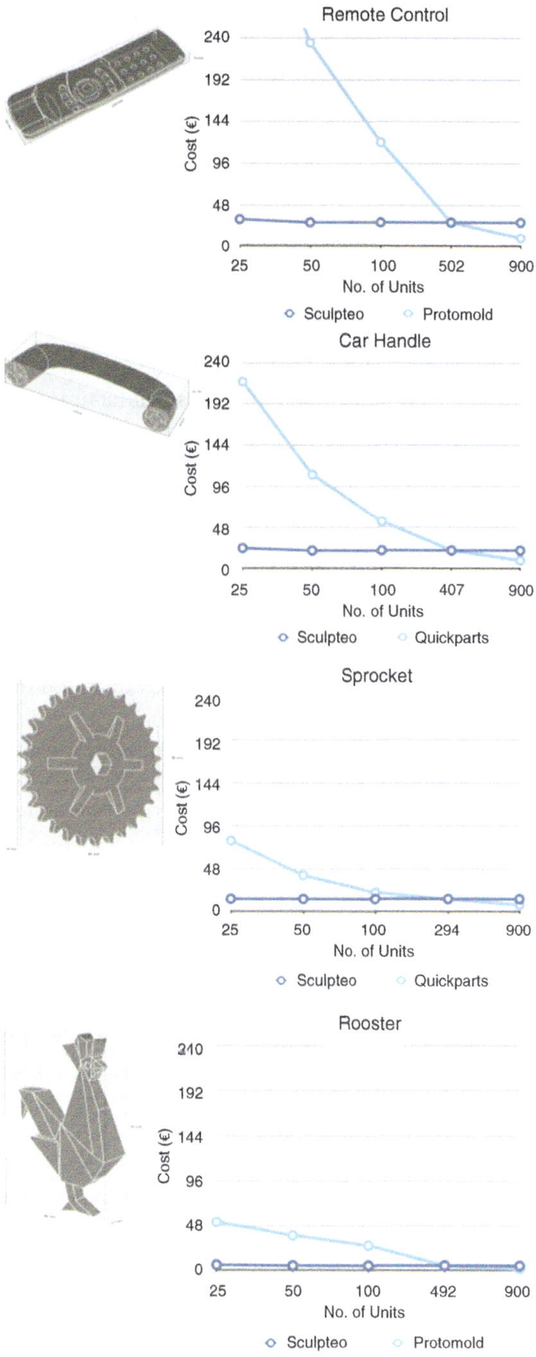

and a 3D printed part consist of different materials, but some of the properties do not differ. However, the 3D printed parts possess various surface finish unlike the injection-molded parts. But surface finish and material properties of the 3D printed part serve its need for any specific products or applications, and the economical manufacturing method for up to 500 unit production runs does not depend on unit size.

Influence of 3D Printing in Industry

In the section of case study, the influence linked with growing part size in both 3D printing and injection molding manufacturing technique is presented and also analyzed.

In order to complete this case study, quotes for 3D printing with a generic, hollow, five-sided box of increasing size as well as injection molding were produced.

By employing "Batch Control" option for selective laser sintering, the 3D printing quotes were produced by the Sculpteo. The cost for the parts in the rapid injection molding was estimated by employing CustomPartNet, which is a free online resource for determining the cost of manufactured parts.

With the help of extensive part gallery, a box was chosen to produce baseline manufacturing cost which determines injection molding. The below table presents the chosen settings for all manufacturing processes.

As mentioned earlier, by employing the CustomPartNet estimation tool, manufacturing cost estimations were produced for part volumes beginning at 10 units and then at intervals of 50 from 50 units to 1000 units. As a result, the data was plotted which enabled comparison of total cost with the number of manufactured units, and the data was obtained from the Sculpteo's batch control option with standard layer thickness. The data reveals that 3D printing is cheap as soon as the size of the object grows. In SLS, the high cost of the largest box offers better results for the cost calculation algorithms. During the SLS process, the machine manufacturing time per unit as well as quantity of material plays an important role such a way if there is an increase in the Z-dimension which is the height of the part in the build volume of the SLS machine, then the component consumes more time for production as well as consumes more material in real quantity as well as manufacturing quantity. Moreover, an increase in production is needed when a component increases in size and few can suit in a single build volume.

However, the rapid injection molding does not experience this challenge providing 3D printing to gain more economic viability in the case of big parts if the production trials are very diminutive, thus making it effective.

There are several advantages for additive manufacturing process in Industry 4.0 when compared with the traditional manufacturing technology. The neural networks automate the manufacturing in almost all areas of sectors [32, 33]. But various parameters tend to limit the usage of AM. In this chapter, the technology behind the AM and traditional manufacturing had been mentioned. With this chapter, a customer could be able to find a suitable technology that fits with the process could be identified. Thus, the working of the manufacturing process has been explained in this chapter. The process variables such as the flow rate and temperature play a major role in this additive manufacturing process. The dimensions and the surface quality should be considered for the proper end product of the material. Although there are certain limitations in the process selection, the best method should be adapted depending on the materials used. New design possibilities and product paradigms along with the new products had bought advancement in the field of AM. To bring the AM into the field, several business plans along with exploring and adopting AM in small and large scale should be made at a best rate.

Conclusion

Although several equipment have been used in the process of AM in Industry 4.0 by quickly varying the parts, it is not necessary that they should have a convenient usage with the enough equality. In case of prototyping, this has an ease usage, but this could not be applicable for educating environment. IoT and AM have its own pros and cons. It's difficult to create a case of one type over the other as a broad recommendation. This recommendation is based on the result and their application purpose. If a product should be created at a faster rate that has the capability to create numerous parts concurrently, then AM could be the best choice of selection. But while considering the surface finish of the material, then AM should be avoided that could shrink the material by size. Hence IoT-based AM could be adapted in this case. It is intellectual to use the combination of both the IoT and AM. We could design and create the parts by means of AM technology in Industry 4.0 with enhanced performance.

References

1. Guan, K., Qin, W., Liu, Y., Yin, X. Q., Peng, C., Lv, M., Sun, Q., & Wu, J. Q. (2016). Evolution of porosity, pore size and permeate flux of ceramic membranes during sintering process. *Journal of Membrane Science, 520*, 166–175.
2. Yang, D. L., Yhang, Y., Song, X. Z., Chen, Y. Z., Shen, Z. Q., & Yang, C. (2016). Effects of sintering temperature and holding time on porosity and shrinkage of glass tubes. *Ceramics International, 42*, 5906–5910.
3. Wilkinson, S. K., Turnbull, S. A., Yan, Z., Stitt, E. H., & Marigo, M. (2017). A parametric evaluation of powder flowability using a freeman rheometer through statistical and sensitivity analysis: A discrete element method (DEM) study. *Computers and Chemical Engineering, 97*, 161–174.
4. Giuntini, D., Chen, I.-W., & Olevsky, E. A. (2016). Sintering shape distortions controlled by interface roughness in powder compositions. *Scripta Materialia, 124*, 38–41.
5. Molla, T. T., Bjork, R., Olevsky, E., Pryds, N., & Frandsen, H. L. (2014). Multi-scale modeling of shape distortions during sintering of bi-layers. *Computational Materials Science, 88*, 28–36.

6. Vasinonta, A., Beuth, J., & Griffith, M. (2008). Process maps for controlling residual stress and melt pool size in laser-based SFF processes. In *Solid freeform fabrication proceedings* (pp. 200–208). Austin: University of Texas at Austin.
7. Gockel, J., Beuth, J., & Taminger, K. (2014). Integrated control of solidification microstructure and melt pool dimensions in electron beam wire feed additive manufacturing of Ti-6Al-4V. *Additive Manufacturing, 1–4*, 119–126.
8. Shen, N., & Chou, K. (2012). *Numerical thermal analysis in electron beam additive manufacturing with preheating effects*. Proceedings of the 23rd annual international solid freeform fabrication symposium, pp. 774–784.
9. Shen, N., & Chou, Y. K. (2012). *Thermal modeling of electron beam additive manufacturing process—powder sintering effects*. Proc. the 7th ASME 2012 International Manufacturing Science and Engineering Conference, pp. 287–295.
10. Ardila, L.C., Garciandia, F., Gonzalez-Diaz, J.B., Alvarez, P., Echeverria, A., Petite, M.M. and Deffley, R. (2014). Ochoa, J. Effect of IN718 recycled powder reuse on properties of parts manufactured by means of selective laser melting. Physics Procedia. 56, 99–107.
11. Prasad, Y. V. R. K., Rao, K. P., & Sasidhar, S. (Eds.). (2015). *Hot working guide: A compendium of processing maps*. ASM International. Second Edition. ISBN: 978-1-62708-091-0. https://www.asminternational.org/home/-/journal_content/56/10192/25536184/PUBLICATION
12. Buffa, G., Campanella, D., Fratini, L., & Micari, F. (2016). AZ31 magnesium alloy recycling through friction stir extrusion process. *International Journal of Material Forming, 9*(5), 613–618.
13. Shao, Y., Tang, T., Li, D., & Peng, Y. (2015). Simulation of the extrusion texture of magnesium alloy AZ31 using crystal plasticity finite element method. *Materials Science Forum, 817*, 538.
14. Wang, X., Wu, M., Ma, W., Lu, Y., & Yuan, S. (2016). Achieving superplasticity in AZ31 magnesium alloy processed by hot extrusion and rolling. *Journal of Materials Engineering and Performance, 25*(1), 64–67.
15. Strásḱa, J., Janeček, M., Čížek, J., Stráský, J., & Hadzima, B. (2014). Microstructure stability of ultra-fine grained magnesium alloy AZ31 processed by extrusion and equal-channel angular pressing (EX–ECAP). *Materials Characterization, 94*, 69–79.
16. Wang, S. C., Zhang, X. H., Hu, X. L., Cai, Z. G., & Peng, B. (2017, March). Study on the hot extrusion forming process of AZ31 magnesium alloy cylindrical shell. In *Advanced materials and energy sustainability: Proceedings of the 2016 international conference on advanced materials and energy sustainability (AMES2016)* (pp. 249–258). World Scientific.
17. Vasinonta, A., Beuth, J., & Griffith, M. (2000). Process maps for controlling residual stress and melt pool size in laser-based SFF processes. In *Solid freeform fabrication proceedings* (pp. 200–208). Austin: University of Texas at Austin.
18. Ardila, L.C., Garciandia, F., Gonzalez-Diaz, J.B., Alvarez, P., Echeverria, A., Petite, M.M. and Deffley, R. (2014). Ochoa, J. Effect of IN718 recycled powder reuse on properties of parts manufactured by means of selective laser melting. Physics Procedia. 56, 99–107.
19. Shao, Y., Tang, T., Li, D., & Peng, Y. (2015, April). Simulation of the extrusion texture of magnesium alloy AZ31 using crystal plasticity finite element method. *Materials Science Forum, 817*, 538.
20. Yu, H., Park, S. H., & You, B. S. (2015). Die angle dependency of microstructural inhomogeneity in an indirect-extruded AZ31 magnesium alloy. *Journal of Materials Processing Technology, 224*, 181–188.
21. Kim, B., Park, C. H., Kim, H. S., You, B. S., & Park, S. S. (2014). Grain refinement and improved tensile properties of Mg–3Al–1Zn alloy processed by low-temperature indirect extrusion. *Scripta Materialia, 76*, 21–24.
22. Tang, W., Huang, S., Li, D., & Peng, Y. (2015). Mechanical anisotropy and deep drawing behaviors of AZ31 magnesium alloy sheets produced by unidirectional and cross rolling. *Journal of Materials Processing Technology, 215*, 320–326.

23. Lu, L., Liu, C., Zhao, J., Zeng, W., & Wang, Z. (2015). Modification of grain refinement and texture in AZ31 Mg alloy by a new plastic deformation method. *Journal of Alloys and Compounds, 628*, 130–134.
24. Yang, Q., Jiang, B., He, J., Song, B., Liu, W., Dong, H., & Pan, F. (2014). Tailoring texture and refining grain of magnesium alloy by differential speed extrusion process. *Materials Science and Engineering: A, 612*, 187–191.
25. Yang, Q., Jiang, B., Zhou, G., Dai, J., & Pan, F. (2014). Influence of an asymmetric shear deformation on microstructure evolution and mechanical behavior of AZ31 magnesium alloy sheet. *Materials Science and Engineering: A, 590*, 440–447.
26. Hu, H. J., Fan, J. Z., Zhai, Z. Y., Wang, H., Li, Y. Y., & Gong, X. B. (2014). Physical fields evolution and microstructures for compound extrusion of AZ31 magnesium alloy. *Russian Journal of Non-Ferrous Metals, 55*(3), 254–262.
27. Pan, F., Wang, Q., Jiang, B., He, J., Chai, Y., & Xu, J. (2016). An effective approach called the composite extrusion to improve the mechanical properties of AZ31 magnesium alloy sheets. *Materials Science and Engineering: A, 655*, 339–345.
28. Drozdenko, D., Bohlen, J., Yi, S., Minárik, P., Chmelík, F., & Dobroň, P. (2016). Investigating a twinning–detwinning process in wrought Mg alloys by the acoustic emission technique. *Acta Materialia, 110*, 103–113.
29. Kecskes, L. J., Hammond, V. H., Eichhorst, M., Herzig, N., & Meyer, L. (2016). *Development of rolling schedules for equal channel angular extrusion (ECAE)-processed AZ31 magnesium alloy*. US Army Research Laboratory Aberdeen Proving Ground United States.
30. Kondoh, K., Hotta, M., Liao, J., Kaneko, K., Fujii, N., Kametani, H., & Koshi, A., Kurimoto Ltd. (2016). *Production method of extrusion billet and production method of magnesium alloy material*. U.S. Patent 9,518,314.
31. Suh, J., Victoria-Hernandez, J., Letzig, D., Golle, R., Yi, S., Bohlen, J., & Volk, W. (2015). Improvement in cold formability of AZ31 magnesium alloy sheets processed by equal channel angular pressing. *Journal of Materials Processing Technology, 217*, 286–293.
32. Suresh, A., Udendhran, R., Balamurgan, M., et al. (2019). A novel internet of things framework integrated with real time monitoring for intelligent healthcare environment. *Springer-Journal of Medical System, 43*, 165. https://doi.org/10.1007/s10916-019-1302-9.
33. Suresh, A., Udendhran, R., & Balamurgan, M. (2019). Hybridized neural network and decision tree based classifier for prognostic decision making in breast cancers. *Springer - Journal of Soft Computing*. https://doi.org/10.1007/s00500-019-04066-4.

Chapter 6
Deep Learning Enabled Smart Industrial Workers Precaution System Using Single Board Computer (SBC)

S. Pradeep Kumar, S. Selvakumari, S. Praveena, and S. Rajiv

Introduction

From the Industrial point of view, the safety measures for the workers are mandatory and must be able to be checked on a routine base. The evaluation of the IIoT-4.0 is given in the following diagram.

In this paper, the single board computer (SBC) is used to solve the mandatory process. The SBC is used to control whether switching options should be enabled or disabled to the power supply of a big machine after verifying that the worker is wearing all the necessary kinds of safety equipment. Deep learning uses image processing to detect the safety objects, such as helmet, gloves, vest coat, etc. The programming language python is used to access the above process with the camera for image processing. In each level of transformation, the input data will be slightly more abstract and a composite mode of representation.

Figure 6.1 shows the evaluation of the IIoT from the Industry 1.0 to Industry 4.0. In the beginning of Industrial 1.0, the mechanization of the weaving loom by steam power was initiated. After that in the 1870s, IoT was widely used for mass production and electrical energy. A huge change occurred in the evaluation of the IoT in 1969 with the marvelous invention of the machine called a computer. Ever since, the

---

S. Pradeep Kumar (✉)
Sri Venkateswara College of Engineering and Technology, Thiruvallur, Tamil Nadu, India

S. Selvakumari
Sriram Engineering College, Chennai, Tamil Nadu, India

S. Praveena
Free Launcher, Thiruvallur, Tamil Nadu, India

S. Rajiv
VIT, Vellore, Tamil Nadu, India

© Springer Nature Switzerland AG 2020

G. R. Kanagachidambaresan et al. (eds.), *Internet of Things for Industry 4.0*,
EAI/Springer Innovations in Communication and Computing,
https://doi.org/10.1007/978-3-030-32530-5_6

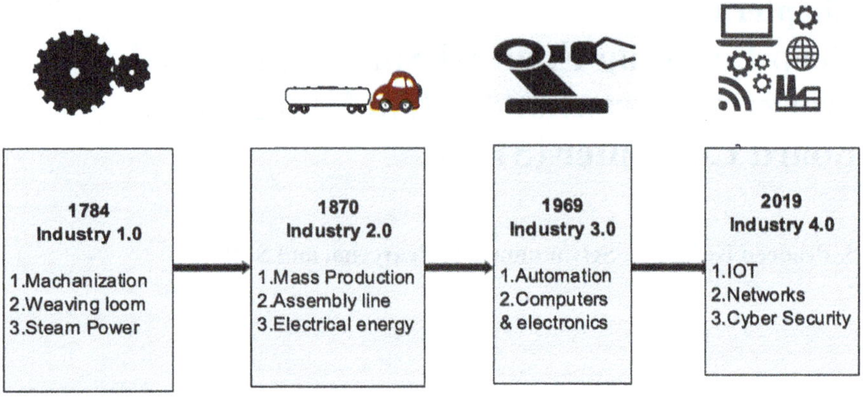

Fig. 6.1 Industrial internet of things 4.0

IoT has played a major role in the industrial process, which includes the communication process with the network and machine with cyber security [1].

This chapter is sorted as follows. Section "Prelims" talks about the framework segments topics used in this project, including those of prelims and architecture of the containing interfaces. Section "Proposed System" gives an outlined description of the proposed method and model of mathematical execution. The graph and the output are condensed in Section "Math Model". Section "Results" gives recommendations for future work, and the chapter is concluded in Section "Conclusion".

Prelims

Image Processing

Imagining processing is the process of changing over ordinary images to digitalized format, and it also carries out certain tasks on the image, such as a request to upgrade a picture or to remove some valuable data from it [2, 3]. Another way of signal agreement to get the input as image, such as video edge or photo yield, might be a picture or specifications about the picture. Typically, the picture preparing framework incorporates treating the image as a few dimensional sign, while previously some type of methods may have been applied to them. The processing of an image commonly includes the following steps:

- Importing the picture using an optical scanner or by computerized picture or by camera.
- Analyzing and controlling the picture, which incorporates information compression and picture upgrade and spotting designs that are not visible to human eyes.

- Output is the last stage wherein the result can be a modified picture or sifted picture that depends on picture examination.

Picture preparation is the procedure of the examination and control of a digitized picture, particularly so as to improve its quality. Observation has the capacity to consequently recognize an individual through facial recognition and may improve modern security strategies. Current progressions in man-made consciousness and machine learning are improving the precision of such systems. Improved component learning and progressed neural system configuration have made article discovery significantly all the more fascinating and productive. The main purposes of the picture preparing include the following: Representation is to watch the items that are not noticeable; Image honing and rebuilding is to make superior pictures; Picture recovery is to look for the picture of intrigue; Estimation of pattern is to quantify different articles in a picture; and Picture Recognition is to recognize the items in a picture.

Deep Learning

Deep learning is a widely used technique, and the "deep" in "deep learning" has lead us to understand the quantity of layers through which the information is changed to achieve the exact output that overcomes the drawbacks of the old algorithm of the machine learning; more accurately called deep learning frameworks. This learning has been applied to the fields of computer vision (CV) of various versions, speech recognition, natural language processing (NLP), audio-recognition (AR), social network filtering (SNF), machine-translation, medical image analysis, and other processes of machine learning concepts [4]. Deep learning designs are regularly built with a ravenous layer-by-layer method of these deliberations and choose which highlights improve execution.

Deep learning strategies hinder highlight designing by making an elucidation of the data into negligible, widely appealing depictions similar to essential parts and derive layered structures that oust abundance in depiction. Profound learning computations can be associated with unsupervised learning assignments [5, 6]. This is a basic bit of leeway in light of the way that unlabeled data are more rich than named data. Cases of profound structures that can be set up in an unsupervised manner are neural history blowers and profound conviction frameworks.

Deep learning is a sort of AI where a model figures out how to perform portrayal assignments straight forwardly from pictures, substance or sound. Profound learning is ordinarily executed utilizing neural framework building. The term profound implies the number of layers in the framework—the more layers, the more profound the framework. Regular neural frameworks contain only a couple of layers, while profound frameworks can have hundreds.

Wireless Sensor Network (WSN)

A Wireless Sensor Network (WSN) is a network of various kinds of devices that communicate the data and information collected from one supervised field through wireless links [7–9]. The Gathered data is transformed through multiple nodes, and the data is connected to other networks with a gateway. The WSN protocol selected may depend upon the requirements of the application. In the WSN, the wireless process may consist of base stations and a number of nodes called wireless sensors. Sensors may vary according to their types, size, speed rate, bandwidth, capacity, and memory. Some of the applications of WSNs are used to measure the environmental physical conditions and to organize the gathered data at some point of location.

TensorFlow

The tool used to write machine learning programs is TensorFlow. TensorFlow is one of the broadly utilized libraries for executing machine learning and different calculations including a huge number of scientific activities. TensorFlow was developed by Google, and it is one of the most popular libraries for Machine Learning. The core component of TensorFlow is the computational graph and Tensors that traverse among all the nodes through edges [10, 11]. TensorFlow is a powerful open-source, artificial intellect library using data flow graphs to build a myriad of models. It is mostly used for classification, perception, understanding, discovering, predicting, and making.

TensorFlow excels at numerical computing, which is critical for deep learning. It has a rich set of application programming interfaces in most major languages and environments needed for deep learning projects. A tensor flow may contain a set of primitive values shaped into an array of any number of dimensions. These monstrous quantities of huge clusters are the reason that GPUs and different processors intended to do skimming point arithmetic exceed expectations at accelerating these calculations and algorithms.

Convolution Neural Network (CNN)

Image classification using CNN is most effectual. Primary and leading, we need a set of imagery. In this case, we take images as our initial training data set [12]. The most common reflection data input parameters are the number of images, image size, number of channels, and the number of levels per pixel.

Figure 6.2 is a clear view of the convolutional neural network explained with the working process and classification models. In the above diagram, the input is taken through the camera, and the processes of convolution and pooling layers are carried

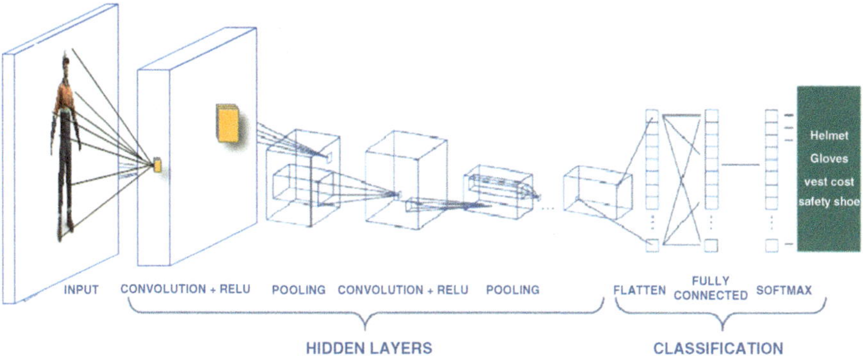

Fig. 6.2 Convolutional Neural Network (CNN)

out and the same processes repeated twice. Some of the algorithm is used to identify the process of safety equipment based on the above method, and with the help of this input, the other process will continue [13]. The output could be compared with deep learning, and based on the output, the SBC will execute the rest of the process.

Proposed System

The working principle of the proposed methodology is in the deep learning concept, and it will verify the occurrence of a helmet and other protection gadgets to ensure the safety of factory workers [13]. The proposed line of attack incorporates the protection aspects confirmation of the labor behavior of cutting and grinding machines. A camera is located before the workbench and connected with the single board computer that may verify the helmet presence through image processing. The verification time of TensorFlow using the Raspberry pi, deep learning method is measured with different subjects and helmets [14]. The large machine will be switched on after confirmation that the worker is wearing a helmet. A camera is located facing the workers and the feed from the camera connected with the single board computer verifies the helmet occurrence through image dispensation (Fig. 6.3).

Deep learning and object detection serve as the main features of this work. The python program code used for object detection is run inside the Raspberry pi type of (single board computer) board to identify objects [14]. The camera sensor and obstacle detection sensors are used to activate the mechanism. Input to the system is from the camera sensor. The single board computer called Raspberry pi is connected with the camera as a USB. Through Wi-Fi, the device is connected to a laptop or any monitor.

Figure 6.4 shows the process of identifying the safety equipment and allowing the smart locking and unlocking of the big critical machine [15]. As per the above diagram, the camera is used to identify the safety equipment of workers if they are not wearing any one of the safety elements the big machine will not switch on and

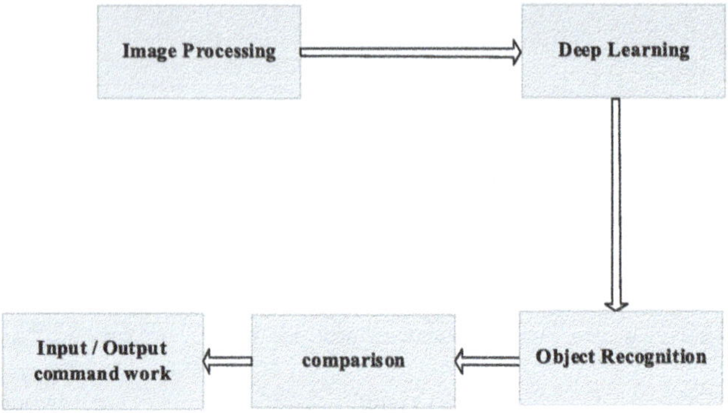

Fig. 6.3 Dataflow diagram

Fig. 6.4 Safety equipment detecting

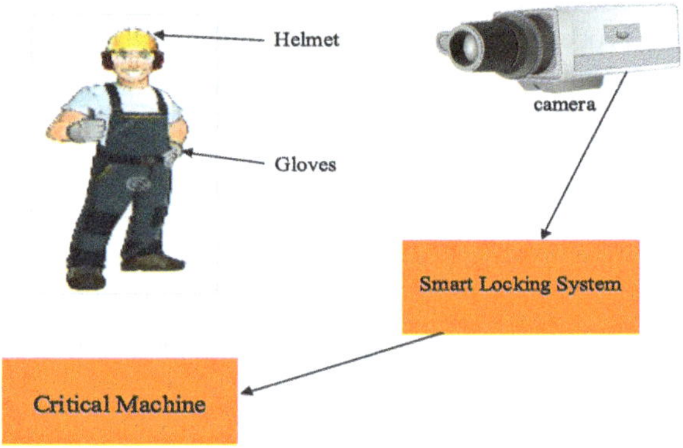

allow work on it [14, 16]. The camera is connected to the single board computer, and that device is trained by the concept of deep learning with the image processing. The Python programming language is used to access the above process.

Math Model

Dissimilar to YOLO, SSD embraces a multi-scale method, which implies that the component maps that are utilized to identify various items are at various scales. Since each element guide is created from convolution results at a similar level, the convolution open fields of the various levels must be different in size [17]. In particular, the receptive fields of a high-level convolution layer are somehow larger than those of lower layers [18–20], and the separated data compared to an abnormal state highlight layer is more abstract. The more abstract the feature extraction information is, the less detailed the information will be; in this manner, SSD discovery is likewise insensitive toward small articles.

The formula for computing the convolution responsive field is the following:

$$S_{\text{CRF}}(i) = \left(S_{\text{CRF}}(i-1)-1\right)N_s + S_f \tag{6.1}$$

where $S_{\text{CRF}}(i)$ is the size of the convolution receptive field of the ith layer, N_s is the step length, and S_f is the size of the filter.

$$a_{\min} = \frac{C_a - \frac{P_b}{2}}{P_{\text{feature}}} P_{\text{img}} = \left(\frac{i+0.5}{|Fk|} - \frac{p_k}{2}\right) P_{\text{img}} \tag{6.2}$$

$$b_{\min} = \frac{C_b - \frac{h_b}{2}}{h_{\text{feature}}} h_{\text{img}} = \left(\frac{j+0.5}{|Fk|} - \frac{h_k}{2}\right) h_{\text{img}} \tag{6.3}$$

$$a_{\max} = \frac{C_a - \frac{P_b}{2}}{P_{\text{feature}}} P_{\text{img}} = \left(\frac{i+0.5}{|Fk|} - \frac{p_k}{2}\right) P_{\text{img}} \tag{6.4}$$

$$b_{\max} = \frac{C_b - \frac{h_b}{2}}{h_{\text{feature}}} h_{\text{img}} = \left(\frac{j+0.5}{|Fk|} - \frac{h_k}{2}\right) h_{\text{img}} \tag{6.5}$$

where (cx, cy) denotes the center coordinates of the default bounding box, h_b is the height of the default bouncing box, w_b is the width of the default jumping box, h_{feature} is the tallness of the component map, w_{feature} is the width of the element map, $|fk|$ is the size in the kth highlight map, w_{img} is the stature of the first picture, w_{img} is the width of the first picture, which is focused at $(i + 0.5|fk|)$, $(j + 0.5|\text{Fsk}|)$ and scaled to a height of h_k and a width of wk in the kth highlight map. In the event that the SSD_300 × 300 model is embraced, with the end goal that the size of the information picture is 300 × 300, the component maps of the model are for the most part created from layers. The sizes of the convolution open field and the mapping area of the default bouncing box on each element map are:

$$\text{L}_{\text{overlap}} = \left(\text{L}_1 \cap \text{L}_2\right) / \left(\text{L}_1 \cup \text{L}_2\right) \tag{6.6}$$

To assess the recovery capacities of the strategies, their capacity to recover questions in each picture was first evaluated, and the normal estimation of these outcomes was then determined [21, 22]. Typically, the recovery capacity is surveyed utilizing the F-measure, which is the weighted normal of the exactness and review and is communicated as follows:

$$\text{F} = \frac{\left(\text{a}^2+1\right)\times \text{P}\times \text{R}}{\text{a}^2 \times \left(\text{P}+\text{R}\right)} \tag{6.7}$$

Results

In this work, we have established a secure working environment for the workers by using the deep learning method and image processing, which help ensure the safety of workers in the industrial environment. Additionally, our work helps us to avoid major accidents in industry and to reduce man power for monitoring, and we are also able to achieve this method at minimum cost [23, 24]. An opportunity for future work can be to extend the process for other wearable safety materials (Fig. 6.5).

Figure 6.6 shows the detection of the safety equipment helmet using our device.

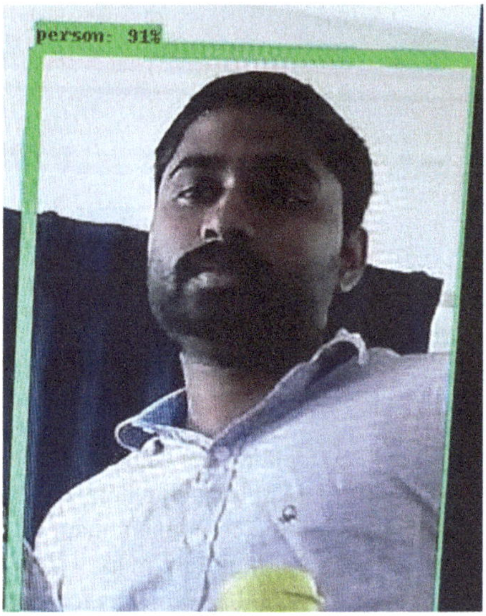

Fig. 6.5 Detecting human and other objects using deep learning

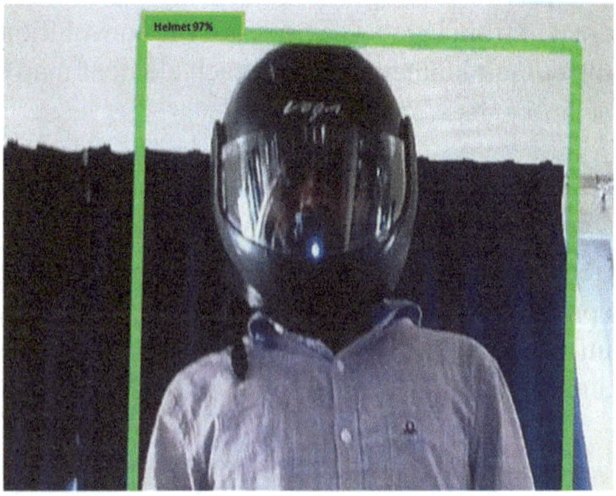

Fig. 6.6 Safety equipment detecting

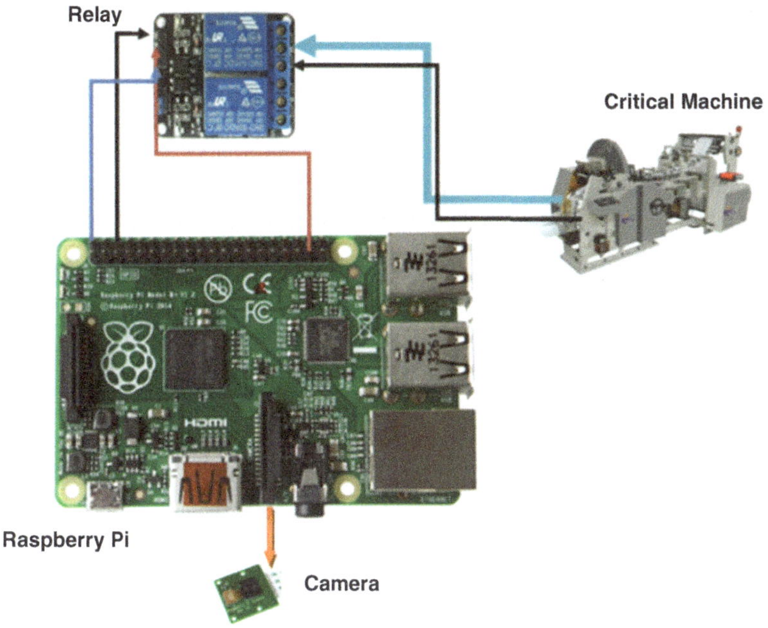

Fig. 6.7 Raspberry pi connected LED instead of critical machine

Figure 6.7 shows the smart locking system of the huge critical machine. After verifying the safety equipment, the device enables the power supply. When the trained device scans the worker who stands in front of the workbench, as soon as the helmet is detected the green light glows. Only when the green light glows is the worker able to power on the machine. If the red light glows, it means the worker is not wearing the safety equipment, and if workers try to start the machine, it will not start.

After ensuring the helmet is worn by the worker, the machine will start, and in the meanwhile, if the worker tries to remove the helmet, the critical machine will shutdown with safety mode. These precautions are for the workers safety so that a major accident can be avoided.

Another expansion of the program is to recognize whether the laborers are wearing their vests; a third augmentation includes adding a program to distinguish a specialist dependent on the framework of a specialist and ordinary leg developments while strolling. Moreover, the utilization of a database of the board for every one of the laborers with the end goal that the program can record every one of the specialists' chronicles in regard to security rule infringement is being investigated. Further examination will proceed until the program can adequately distinguish all the PPEs (vests, boots, globes, and so forth). It will be simpler to recognize vests when utilizing this program than to distinguish boots and gloves on the grounds that the cameras by and by cannot accurately catch the state of the boots and gloves on account of their small size contrasted with vests.

Conclusion

In this work, we have established a secure working environment for the workers by using the deep learning method and image processing, which help to ensure the safety of workers in the industrial environment. Additionally, our work helps us to avoid major accidents in industry and to reduce man power for monitoring, and we are also able to achieve this method at minimum cost. In future, the work can be extended for other wearable safety materials.

References

1. Din, S., Paul, A., Ahmad, A., Gupta, B. B., & Rho, S. (2018). Service orchestration of optimizing continuous features in industrial surveillance using big data based fog-enabled internet of things. *IEEE Access, 6*, 21582–21591.
2. Krüger, J., Lien, T. K., & Verl, A. (2009). Cooperation of human and machines in assembly lines. *CIRP Annals, 58*(2), 628–646. [2].
3. Haddadin, S., Albu-Schäffer, A., & Hirzinger, G. (2009). Requirements for safe robots: Measurements, analysis and new insights. *The International Journal of Robotics Research, 28*(11–12), 1507–1527.
4. Asadi, E., Li, B., & Chen, I. M. (2018). Pictobot: A cooperative painting robot for interior finishing of industrial developments. *IEEE Robotics & Automation Magazine, 25*(2), 82–94.
5. Krizhevsky, A., Sutskever, I., & Hinton, G. E. (2012). Imagenet classification with deep convolutional neural networks. In *Advances in neural information processing systems* (pp. 1097–1105).
6. Xia, F., Campi, F., & Bahreyni, B. (2018). Tri-mode capacitive proximity detection towards improved safety in industrial robotics. *IEEE Sensors Journal, 18*(12), 5058–5066.
7. Martin, A., & Voix, J. (2018). In-ear audio wearable: Measurement of heart and breathing rates for health and safety monitoring. *IEEE Transactions on Biomedical Engineering, 65*(6), 1256–1263.
8. Lazzerini, B., & Pistolesi, F. (2018). Multiobjective personnel assignment exploiting workers' sensitivity to risk. *IEEE Transactions on Systems, Man, and Cybernetics: Systems, 48*(8), 1267–1282.
9. Lazzerini, B., & Pistolesi, F. (2018). An integrated optimization system for safe job assignment based on human factors and behavior. *IEEE Systems Journal, 12*(2), 1158–1169.
10. Tsukada, K., Tomioka, T., Wakabayashi, S., Sakai, K., & Kiwa, T. (2018). Magnetic detection of steel corrosion at a buried position near the ground level using a magnetic resistance sensor. *IEEE Transactions on Magnetics, 54*(11), 1–4.
11. Rybczyński, A., Wolska, A., Wisełka, M., Matusiak, J., & Pfeifer, T. (2018). Welding arc ignition and photobiological hazard evaluation. In *2018 IEEE International Conference on Environment and Electrical Engineering and 2018 IEEE Industrial and Commercial Power Systems Europe (EEEIC/I&CPS Europe)* (pp. 1–6). IEEE.
12. Doan, D. R. (2018). Unrecognized hazards [electrical safety]. *IEEE Industry Applications Magazine, 24*(5), 5–5.
13. Doan, D. (2017). Equipment testing safety [electrical safety]. *IEEE Industry Applications Magazine, 23*(4), 6–7.
14. Doan, D. (2017). Dangers of distractions [electrical safety]. *IEEE Industry Applications Magazine, 23*(3), 5–72.

15. Spang, R. J. (2017). Integrated safety management: Creating an all-inclusive electrical safety program. *IEEE Industry Applications Magazine, 23*(3), 64–70.
16. Mayer, S., Hodges, J., Yu, D., Kritzler, M., & Michahelles, F. (2017). An open semantic framework for the industrial Internet of Things. *IEEE Intelligent Systems, 32*(1), 96–101.
17. Doan, D. (2017). Workplace hazards for the electrical worker' [electrical safety]. *IEEE Industry Applications Magazine, 23*(2), 6–6.
18. Krizhevsky, A., Sutskever, I., & Hinton, G. E. (2012). Imagenet classification with deep convolutional neural networks. In *Advances in neural information processing systems*, pp. 1097–1105.
19. Lawrence, S., Giles, C. L., Tsoi, A. C., & Back, A. D. (1997). Face recognition: A convolutional neural-network approach. *IEEE Transactions on Neural Networks, 8*(1), 98–113.
20. Chen, C. F., & Hsiao, C. H. (1997). Haar wavelet method for solving lumped and distributed-parameter systems. *IEE Proceedings-Control Theory and Applications, 144*(1), 87–94.
21. Henriques, V., & Malekian, R. (2016). Mine safety system using wireless sensor network. *IEEE Access, 4*, 3511–3521.
22. Khan, M., Silva, B. N., & Han, K. (2016). Internet of things based energy aware smart home control system. *IEEE Access, 4*, 7556–7566.
23. Li W., Zhao R., Xiao T., & Wang X. (2014). DeepReID: Deep filter pairing neuralnetworkforpersonre-identification. In *Proceedings of the IEEE conference on computer vision and pattern recognition* (pp. 152–159).
24. Luo, P., Tian, Y., Wang, X., & Tang, X. (2014). Switchable deep network for pedestrian detection. In *Proceedings of the IEEE Conference on Computer Vision and Pattern Recognition* (pp. 899–906).

Chapter 7
A Novel Design Augmentation of Bio-Inspired Artificial Immune Technique in Securing Internet of Things (IOT)

G. Usha, P. Madhavan, and M. V. Ranjith Kumar

Introduction

Various germs, viruses, and other pathogens are more harmful to the human immune system. However, the human immune system has the ability to protect against various attacks and threats to keep the body healthy. Various types of basic definitions are existing to understand the concepts of self, nonself, and antigen identification [1, 2]. The artificial immune systems (AIS) [3] are a type of computational intelligent techniques. The AIS are based on rule-based machine-learning techniques that are based on human immune systems. These artificial immune systems [3] are used in various types of problem-solving techniques. These artificial immune systems (AIS) are new bio-inspired model, which are used to solve information security. AIS techniques [4] are akin to other problem-solving techniques such as genetic algorithms, neural networks, evolutionary algorithm, and swarm intelligence techniques. AIS techniques [5] consist of unique features such as distributed, dynamic in nature, self-learning, self-adapting, and self-organizing in nature. The main unique features for using AIS technique is that it uses an easy method to identify patterns. Various types of immune algorithms [3, 6] are existing to security issues of immune systems. These are shown in Fig. 7.1.

The bio-inspired algorithm has been playing a vital role in the AIS. The simple taxonomy of algorithms are depicted in Fig. 7.1.

The negative selection algorithms (NSA) has to be accessed by hypothesis-resistant techniques. These can be included with the self and nonself models to create and identify arbitrary and counteracting elements. Self-accessing techniques managed by the T cell can be measured at the development stage.

G. Usha (✉) · P. Madhavan · M. V. R. Kumar
SRMIST, School of Computing, Kattanlathur, Kanchipuram District, India
e-mail: ushag@srmist.edu.in

© Springer Nature Switzerland AG 2020

G. R. Kanagachidambaresan et al. (eds.), *Internet of Things for Industry 4.0*, EAI/Springer Innovations in Communication and Computing,
https://doi.org/10.1007/978-3-030-32530-5_7

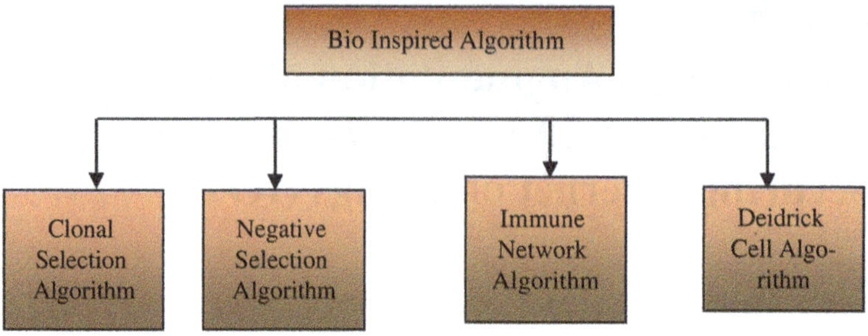

Fig. 7.1 Types of bio-inspired algorithm

In a negative selection algorithm, clonal choice techniques are used to assess the flexibility to wireless sensor network IDS plan. The resistant area of this work attempts to recreate fragmentations of the hypothesis resistance techniques in a wireless sensor network SN and that endeavors to determine an IDS scheme. This is completed utilizing the properties of negative determination.

The negative determination is utilized in hypothesis techniques to show intrusions in sensor networks. Negative choice is commonly not favored in planning intrusions for sensor networks, because of the quantity of finders it creates. Nevertheless, this work uses highlights, for example, time span, time interim between exchanges, information sent, and so on.

The negative selection algorithms with self and nonself mechanisms are used for creating substances. To toggle it, danger theory can be used that rotates around risk in dendritic cells when an interruption or an oddity is recognized.

The blue penciling techniques with NSA attempts to create various arbitrary parallel strings. The strings are coordinated as self, and the remaining are named as an identifier set. This blue penciling is utilized to coordinate the types in the system when the identification stage is named as an irregularity cell.

Negative Selection Algorithm will undoubtedly have a huge number of contrast data because it does not have a unified hazard gauging instrument and has a similarly conflicting memory locator component. Similarly, negative determination is not a thickly populated condition or a system with substantial traffic. The traffic builds in a negative choice algorithm and the quantity of locator sets increments. Since the indicator set is arbitrarily created, it builds the idleness and turns out to be computationally invalid.

The negative determination calculation is not reasonable for thick situations, as it will create the high false negatives. High false positives cannot adapt up to changing elements of the framework with respect to time. It cannot adapt up to evolving elements, memory detectors, and faster response time,

Clonal selection algorithm is another type of algorithm that is additionally used to provide safe framework. The clonal selection is fundamentally utilized to negative selection for the recognition system, which is dependent on intrinsic insusceptible properties utilizing an unsupervised AI approach. This calculation depends on distinguishing self and nonself designs, which is the center of negative determination. The clonal selection can be utilized for negative determination along with clonal choice administrator. The system conventions, for example, TCP and UDP are grouped in the wake of doling out a quality worth and a lot of locator genotypes are determined.

The clonal choice is typically utilized to a negative choice administrator or a liking adding machine, which is utilized to ascertain the fondness between the counteracting agent and the antigen. The locator set shaped from the negative determination calculation is cloned, and a lot of indistinguishable clones are created. These clones, in any case, are not by any stretch of the imagination indistinguishable. Hyper-transformation guarantees that the clones produced are of different types of the parent clone; however, they are of a similar string size. This has the impact of making the finder set bigger and increasingly complete. The clone choice calculations are not appropriate for thick conditions, low false positives, and cannot adapt up to evolving elements, memory identifiers, and faster reaction time,

The fundamental highlights of immune network algorithm are programmed assurance of the populace estimate and blend of neighborhood with worldwide hunt. An insusceptible calculation, named CLONALG, was created to perform pattern acknowledgment and advancement. The creators exhibited exactly that this calculation is fit for learning a lot of info designs by choosing, replicating, and changing a lot of "counterfeit safe cells." The creators demonstrated the reasonableness of the calculation for multimodal search and introduced experimental outcomes where it could beat a wellness sharing methodology.

The artificial immune system engaged with CLONALG is likewise found in a developmental calculation, enabling it to be described as a transformative calculation motivated in the insusceptible framework. It is noted that there is a significant calculated contrast between the clonal choice calculation and a transformative calculation. In the previous, the hypothesis of advancement is utilized to clarify the conduct of the framework, while in the last it motivated its improvement. Various fascinating highlights of pick AINet are listed as follows.

1. It shows a deterministic and elitist determination system for each clone.
2. The cardinality of the populace is consequently controlled by the concealment.
3. The quantity of newcomers' increments increase as the populace increments in size. This is in such a case that the populace is consistently expanding in size, which means that the issue has numerous nearby optima and the more optima it finds, the more it is equipped for finding.
4. No encoding of the people of the populace is required.

There exist various contrasts between pick AINet and CLONALG. CLONALG encodes the people of the populace utilizing parallel strings, whereas AINet depends on genuine esteemed vectors. The select AINet incorporates the communication of the system cells with the earth (wellness) and with one another (partiality), permitting the dynamic control of the populace estimate. In the pick AINet case, newcomers are just permitted to enter the populace after the present cells.

In CLONALG, the partiality proportionate transformation depends on a control procedure where short blasts of high transformation rates are trailed by some breathing periods. Interestingly, AINet pursues a Gaussian transformation that is conversely relative to the standardized wellness of each parent cell. Select AINet additionally introduces various likenesses with the advancement techniques (ES).

These algorithms are used to explain the mammalian adaptive immune system behavior. Internet of Things (IoT) are used in various types of smart devices such as smartphone and various types of sensing devices in order to transfer data locally and to exchange various types of information [6]. Nowadays, the IoT technology is used in mobile applications, entertainment, and other new context in order to realize the better user interface, which provides cheaper, reliable, and robust communication techniques. The important and most challenging concept in IoT is the design of miniature things provided with security and privacy.

Security is a very important task for increasing the number of connected devices in IoT. Every organization transfers the data faster than the content of the data. Security plays a major role in data transfer between organizations. Data security and privacy are very important issues in IoT to secure data. If a user needs to access unauthorized data, it violates the security property of the complete network. In order to use IoT in a broader range, several issues such as security and privacy are most important aspects that need to be solved.

In order to provide authentication and access control, the security and privacy control techniques need to be addressed. IoT applications need to be prevented from security and privacy problems. The security solutions need to prevent illegitimate access and should enable legitimate users to access resources in an authorized manner [7].

While understanding security and privacy of the IoT application, we need to address the intrusion detection and prevention systems. The main aim of the intrusion detection system is to distinguish between self and nonself patterns. Intrusion detection techniques play major role in detecting and securing a network from malicious activity [7]. The IoT technique faces various types of limitations in intrusion detection technique [3]. The intrusion detection systems lack to provide standardized protocol. The resources for computation are minimal. In order to overcome the shortcomings, we provide an IoT-based design technique, which is used to detect and prevent attacks in distributed environment with low computational and self-organization complexity. The proposed technique also provides a secured optimization technique that provides security in IoT [8]. The organization of the work is structured as follows. Section "Theoretical Background" describes the theoretical background emphasizing negative selection and honeybee optimization algorithm [5]. Section "Proposed System Architecture" describes the organization of the proposed intrusion detection design technique [9]. Finally, this work is concluded with the conclusion and with future work.

Theoretical Background

Various techniques are available to classify self/nonself classification methodologies. The techniques are based on statistical techniques or signature-based technique. The signature-based [3] technique needs predefined attack patterns to properly identify attack. Additionally signature-based technique also requires large amount of data. These types of signature-based attack techniques cannot detect new type of attack patterns [1]. The statistical technique relies on the data with statistical value. These techniques are not suitable for IoT intrusion detection technique [10]. In this work, we have proposed a dynamic optimized security technique to detect and prevent attacks in IoT.

Neural Network–Based Negative Selection Algorithm

Much research work has been carried out in order to use neural networks in IoT platform. In this approach, we use simple perceptron-based neural network technique, which uses supervised learning with binary classification [1]. This binary classification algorithm correctly classifies the set into linear separable sets. The input x is mapped to a single binary value. The input function f(x) is given as

$$f(x) = \begin{cases} 1 \text{ if } w.x + b > 0 \\ 0 \text{ otherwise} \end{cases} \tag{7.1}$$

"w" is known as weight vector, "b" is known as bias, and "$w.x$" is given in Eq. (7.2)

$$\sum_i^m w_i \cdot x_i \tag{7.2}$$

Here "m" is number of inputs and "." is known as the dot product. Similarly, a large number of different classes are solved using feed-forward neural network. This technique solves a number of neurons with different layers using the activation function and the number of hidden layers. In order to simplify the proposed technique [1, 2], it is assumed that the input does not have any classification errors. In a negative selection algorithm, it is inspired by self-nonself discrimination of mammalian immune systems [8].

This algorithm suggests the principle of self-nonself discrimination approach. This technique is followed by discriminating anomaly and self-pattern detection techniques. As discussed earlier, the input is classified as two sets. One is a normal set denoted as N, and another set is anomaly data denoted as A. The complete dataset is represented as C, where

$$C = NA \tag{7.3}$$

Next, we analyzed the working of honeybee algorithm and the purpose of using it in the proposed work [11].

In Eq. (7.3), the NA entity is known as normal and in order to classify NA as attack or normal set, the following classification equation is used:

$$D = (f, M) \tag{7.4}$$

where "f" is known as classification function. "M" consists of multiple detector sets. In order to further provide secured optimization technique, we use honeybee algorithm to optimize.

Honeybee Algorithm

The honey bee algorithm is an improvised technique for artificial immune systems for intrusion detection. The honey bee must fly for long distances and run around to different sources flowers so as to gather spore or nectar. In such a state of affairs, the food supply has an additional variety of nectar. During a colony, the scout bees start the method of hunting and search for the patch in a colony. The information containing food supply is not available for scout bees.

Toney honey bee calculation is primarily based on bunching structure teams. A hub is chosen as cluster head obsessed on hub degree, neighbor's conduct, skillfulness course, mutability speed, and remaining vitality. In addition, the planned strategy evolved from the bumble bees and provides practiced and stable bunch development.

The honey bee calculation is to be done for the process of collecting food information. In this technique, each procedure has its terribly having their own qualities and confinements. During this work, the honey bee calculation is done with the calculation of to the grouping issue. The subsequent space controls the essential system of the honeybee calculation.

The honey bee begins by flying for the rationale for scavenging food. The honey bees gather food dirt or nectar from varied sources. On the off probability that the sustenance supply contains a lot of nectar, a lot of honey bees can visit that site for his or her assignment to assemble a lot of nectar. The scout honey bee begins the method toward rummaging by haphazardly ransacking through the fix in an exceedingly settlement. The scout has no earlier information of nourishment sources.

The scout could be a type of honey bee that appears for nourishment source and provides steering to operating drones. With that, the settlement announces a neighborhood of its public as a scout honey bee for investigation throughout the gathering season. The honey bee assesses various fix characteristics that supported the presence of sugar or nectar within the patch. The honey bee plays out a form of move referred to as the "waggle move." The presence of nectar and therefore the heading of the sustenance supply area unit are then determined.

The bearing of sustenance is speculated smitten by the honey bee move therein specific heading. The speed of move demonstrates the presence of nectar within the sustenance supply. Moderate move exhibits that the sustenance supply contains less nectar and therefore the different method around. This move is very important for correspondence between the state members. Through on these lines, the operating drones sent to various sustenance hotspots for nectar accumulation with none guide, control, or another steering. The operating drones look out for the move floor for the waggle move of scout honey bees.

The operating drones begin their journey smitten by the sort of move seen on the move floor. On the off probability, if move is fast, a lot of operating drones area unit sent there sustenance hotspot for nectar gathering. Also, the state accumulates a lot of nectar in an exceedingly expert method. Succeeding waggle move is basic for the scout, once coming back to the nectar. If there is still enough dirt within the sustenance supply, the move is promoted, and therefore the manual laborer honey bees area unit sent to the sustenance supply. The honey bee calculation has been widely utilized within the writing for bunching in information to the mining.

The patch qualities' area unit are determined by the quantity of sugar or nectar within the patch. Whenever a food supply is found, the bees perform a dance referred to as "waggle dance." Based on the gap and the quantity of food supply, the direction of the food supply is calculated. Thus, the bee dance plays a vital role in judging the gap of the food supply. The dance speed is employed to understand the quantity of nectar within the food supply. Once the dance is slow, it imposes that the supply has terribly less quantity of food supply. This dance plays a major role in communicating between the members of the colony.

At the same time, the other bees (worker bees are selected for detecting different food sources)collect the nectar from other food sources. This is achieved with the help of the scout bees' dance. As mentioned before, if the dance is faster, the worker bees collect the food from the nectar. These types of algorithms are very useful in path finding problems in data mining. In our technique, we use this honeybee algorithm [11] to search for the secured optimized resource in IoT.

In honeybee calculation, mostly bunch honey bee calculation could be a multi-target sweetening procedure propelled from the looking conduct of nectar honey bees. It is connected to varied streamlining problems, such as, the voyaging sales rep issue, longest traditional subsequence, and framework chain duplication. In our planned arrangement, the bumble bee calculation is utilized to isolate the system into bunches. We expect numerous parameters during cluster development, such as, hub movable, hub degree, hub remaining vitality, and neighbor quality. Our planned arrangement begins with causing honeybees within the hunt area. In HBAC (Honey bee calculation), the honey bees square is organized into three distinct gatherings: scout honey bees, footer honey bees, and utilized honey bees.

The worker honey bees handle investigation on the nourishment sources superior to the previous one. The spectator honey bee appear out for the move floor for the waggle move of the utilized honey bees. The spectator honey bee chooses the sustenance supply addicted to the type of move that utilized honey bee performs. The sort of move demonstrates the heading of sustenance even as the presence of nectar therein. The honey bees square measures the operating drones that play out the nectar accumulation.

During this work, the looking conduct of honey bees calculations (HBAC) is employed to settle on the choosing of cluster heads. Bunches square measure formed addicted to the selected cluster heads. The task of CH is distributed to totally different hubs that have additional nectar than the previous one.

The planned HBAC calculation works in two stages. The primary one is the bunch arrangement and the second is cluster maintenance. In this stage, the honey bee province is isolated into three sets; the principal 100% of the world is thought as scout honey bees. The second forty-fifth is employed honey bees, whereas the remaining forty-fifth is labeled as spectator honey bees. The scout honey bees speak to all or any out bunches within the system. The hubs within the system square measure isolated into bunches utilizing HBAC calculation. The operating system of arrangement stage is expressed in rule one.

1. Initialize the preparation tests
2. Produce the opening populace Xi, $i = 1, \ldots, N$ utilizing condition (7.1)
3. Compute the wellness of the sustenance sources Q(Neci) utilizing condition (7.6)
4. Do (while set of utilized honey bee)
   (i) Make new nourishment sources (arrangements) FSi utilizing condition (7.5)
   (ii) Figure the worth Q(Neci)
   (iii) Apply ravenous choice procedure
   (iv) End
5. Figure the likelihood pi for the nourishment source Xi by condition (7.4)
6. Do (while set of spectator honey bee)
   (i) Select an answer Xi relying upon pi
   (ii) Produce sustenance sources (arrangements) FSi
   (iii) Decide the worth Q(Neci)
   (iv) Apply ravenous choice procedure
   (v) End
7. On the off chance that there is a dismissed sustenance hotspot for the scout honey bee
8. Substitute it with new sustenance source which will be unpredictable shaped by condition (7.1)

Along these, spectator honeybees are equivalent number to utilize honey bees. The calculation starts and focuses (called introductory CHs) in the search space. The CHs are chosen arbitrarily and wellness is assessed. The wellness of a CH depends on the measure of nectar. The CH having more nectar is a decent contender for the CH job when contrasted with those having less nectar. The measure of nectar in a hub relies upon a few parameters, Hub vitality (Hnode). In this parameter, the hub having most extreme power is designated as a solid contender for CH. Also, the hub having medium power is known as an ideal hub. The hub having a base power is known as a frail hub.

Neighbor quality (Nnode) is the neighbor with one bounce separation and viewed as a decent neighbor. The second neighbor that jumps away is viewed as the moderate neighbor, and neighbors far away are viewed as an awful neighbor.

Hub degree (Anode) is the hubs having a degree over 10% are viewed as incredible. A hub having a degree somewhere in the range of 5% and 10% is viewed as a decent hub. The hub with a less degree is known as a terrible hub.

Hub portability (Pnode) is the hub with relative versatility and is known as recom-repaired. Correspondingly, the hub with various versatilities and possessing a similar course is known as incompletely prescribed. The hubs with various probabilities and varying bearing are set apart as not suggested for CH.

The measure of nectar in a hub xi can be determined by

$$\mathrm{hi} = \mathrm{H\,node} + \mathrm{Nnode} + \mathrm{Anode} + \mathrm{Mnode} \tag{7.5}$$

where hi is that the hub to be investigated. The capability of a hub to show into the CH is set (streamlined) in order that the geometer separation of every CH to alternative CH need to be round the same. The nectar sources are processed (optimized) in order that the geometer separation of each nourishment space to alternative sustenance are – as need to be roughly the equivalent. Here, there is the difficulty of gathering *N* specially appointed system hubs into a *K* range of noncovering bunches.

The problem of grouping network nodes into a *K* range of nonoverlapping clusters is measured. The pursuit system of the scout, spectator, and used honey bees is rehashed to deliver another people of the CHs (arrangements). The memory of the used honey bees that stores CHs (solutions) have visual knowledge and also the tests performed for checking the character of the new CHs (novel arrangements).

Q(Node)is extremely the quantity of nectar at hub "*i*." The passer-by honey bees watch the moving procedure of the used honey bee and disentangle to go to the CH hub Xj with probability pi determined as

$$\mathrm{pi} = \mathrm{Q}(\mathrm{Neci}) / \sum_{\mathrm{ik}=1}^{\mathrm{fs}} \mathrm{F}(\mathrm{Neck}) \tag{7.6}$$

$$\mathrm{FSi}(x+1) = \mathrm{FSi}(x) + \mathrm{aij} \tag{7.7}$$

Here, when the measure of accessible nectar determined for novel CH is better as thought about than the past, the honey bee memorizes the new nectar sum and overlooks the past. If not, the area of the prior CH is kept up in her memory with no modification.

The utilized honey bees share the amount of nectar of various CHs and their directions with different honey bees on the moving floor when they come back to the hive in the wake of finishing the pursuit procedure. At the point, when an onlooker honey bee watches the move of honey bees that are moving on the move floor, she breaks down the nectar sum by watching the sort of move that utilized honey bees perform.

The honey bees check the number of nectar in human positions. The new location is preserved and overlooks this position place away in her memory (alter the position cluster in their memory) subject to the wellbeing of the hopeful's position. The probable rationalization to the bunching issue is that the space of the nectar positions and also the presence of nectar (its wellness) speak to its quality connected with the conceivable arrangement and may be determined by the condition

$$\mathrm{Q}(\mathrm{Neci}) = (1) / (1 + \mathrm{cfi}) \tag{7.8}$$

Once manufacturing of food, every candidate food location is compared with the prevailing (old) food position. The previous one is preserved as long as the nectar quantity of candidate food position is a smaller amount than the previous one which is given by (7.9).

$$Y(x) = NO(Y(x\_1)) \tag{7.9}$$

The choice between the previous and candidate answer is disbursed with the greedy mechanism. The management parameters employed in this approach area unit a most range of rounds and also the most CHs (clusters). The search method is meant in such the way that each of the exploration and also the exploitation of honey bees may be disbursed conjointly.

Cluster Maintenance and Communication
The cluster maintenance and communication continue to play a vital role in the security issues. With that, the bunches area unit formed in cluster arrangement stage, the subsequent stage is to primary teams if any adjustment in skilfulness and vitality exhaustion of a hub happens. The bunches may be effectively primary sustained, if the hubs and their neighbors area unit characterized.

The hub metal within the cluster that starts honey bees (short scrounging extent honey bee) that may visit hubs within the system instating as of hub metal. The quantity of hubs within the bunch is resembling its size. The set CH is in addition characterized that stores the information of CHs after the effective bunch development.

Proposed System Architecture

Data Preprocessing and Data Analysis
Initially the input is given which is classified as set of self patterns and set of nonself patterns [12]. The number of hidden layers is calculated, the number of neurons in each layer is also computed, and the rules are set for training. The abovementioned algorithm is used to generate the detector technique [12]. For neural network training, the self patterns are recognized and for each pattern the corresponding bits are matched. In this, the number of negative detector set [8] is calculated as the size of the negative set. From this, the output is calculated as set of detectors, which are capable of classifying nonself patterns.

The process of naïve detector [11] is that it generates the values randomly. After that, the affinity value of naïve detector value is computed with each member in the self-set. Based on the threshold if the naïve detector value does not match any element in self-set that value is added to negative detector set. The abovementioned steps are used with the help of single-layer feed-forward neural network [12]. This method is simple, fast and used to classify the technique as self and nonself classification. In this step, finally we obtain the trained data.

Figure 7.2 represents the proposed system architecture in detail. Now we discuss the design process in detail.

Data Optimization
After identifying the secured data, the information need to be traveled in optimized manner. Hence, in order to optimize the proposed work, we use honeybee algorithm [11]. In this technique, both local and global search techniques are used. After getting the trained data, we are generating the initial population, which consists of size

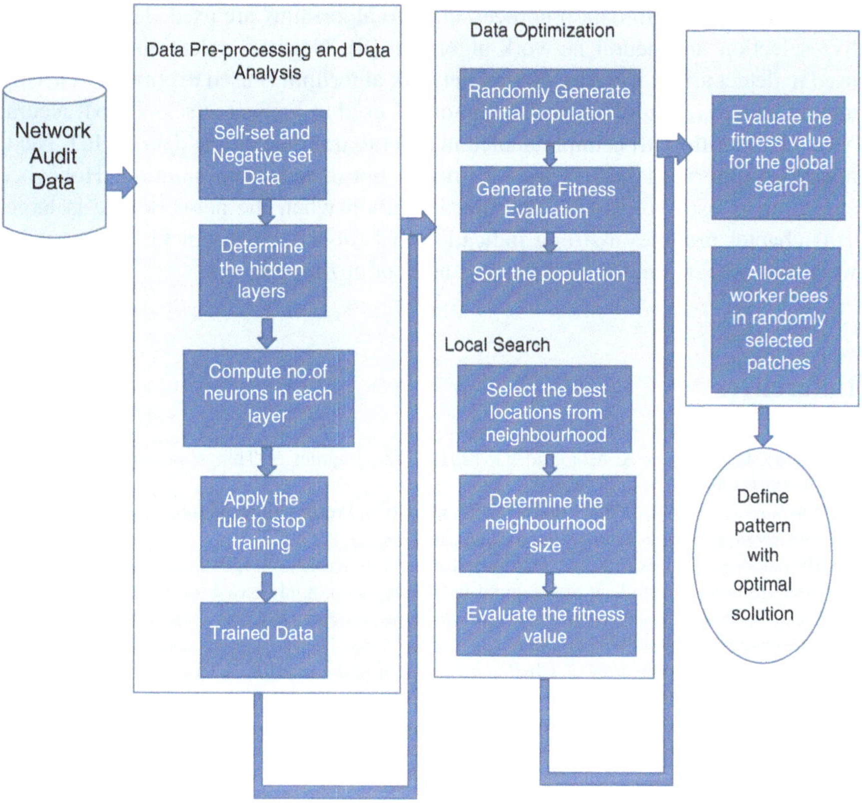

Fig. 7.2 Proposed system architecture

n for worker bees [11]. The fitness evaluation function is known as the initial population. The input values are calculated from maximum to minimum. First in optimization, local search technique is used. In this local search technique, the worker bees calculate the neighborhood size based on the neighborhood search [11].

Finally, the fitness value is allotted based on best location. In the global search technique, the evaluation value is calculated and the values are allotted randomly to worker bees. In this way, the optimized path is selected. The abovementioned process is used to transfer data from one node to another node in IoT in a secured manner. The proposed technique is used to transfer data securely and in optimized way.

Conclusion

In this chapter, we proposed a novel design architecture to detect and prevent attacks in IoT. The proposed design framework is used to detect and prevent attacks in IoT. The proposed work consists of data preprocessing and data analysis and data optimization steps. In data preprocessing two algorithms are used. They are negative selection and neural network algorithm [8]. Negative selection algorithm is used to detect attack patterns. Neural network algorithm is used to correctly classify unseen patterns. Neural network technique is used to prevent attacks in IoT. Neural network algorithm [8] is implemented in order to train the online data for IoT. Next, honeybee algorithm [11] is used to detect optimized route solution. Honeybee algorithm is used to provide optimized solution when the network size is large. This chapter provides a strong indication that for IoT environment, the proposed work will outperform for both security and optimization issues.

References

1. Atzori, L., Iera, A., & Morabito, G. (2010). The Internet of Things: A survey. *Computer Network Journal, 54*, 2787–2805.
2. Kawamoto, Y., et al. (2014). Internet of Things (IoT): Present Stateand future prospects. *IEICE Transactions on Information and Systems, E97-D*(10), 2568–2575.
3. Hosseinpour, F., Bakar, K. A., Hardoroudi, A. H., & Kazazi, N. (2010, Nov). Survey on artificial immune system as a bio-inspired technique for anomaly based intrusion detection systems. In *Intelligent networking and collaborative systems (INCOS), 2010 2nd international conference on* (pp. 323–324). IEEE.
4. Stout, W., & Urias, V. (2016). Challenges to securing the Internet of Things. In ICCST.
5. Jing Liu, Yang Xiao., & Philip Chen, C. L. (2012). *Authentication and access control in the Internet of Things*. 32nd international conference on distributed computing systems workshops.
6. Kolias, C., Kambourakis, G., & Maragoudakis, M. (2011). Swarm intelligence in intrusion detection: A survey. *Computers & Security, 30*(8), 625–642.
7. Alaparthy, V. T., & Morgera, S. D. (2018). A multi-level intrusion detection system for wireless sensor networks based on immune theory. *IEEE Access, 6*, 47364–47373.
8. Pamukov, M. E., Poulkov, V. K., & Shterev, V. A. (2019). Negative selection and neuralnetwork based algorithm for intrusion detection in IoT. In *Proceedings of the 2018 41st international conference on Telecommnications and Siganl Processing (tsp)* (pp. 1–5). IEEE.
9. Shi, Y., Li, T., Li, R., Peng, X., & Tang, P. (2017). An immunity-based IOT environment security situation awareness mode. *Journal of Computer and Communications, 5*, 182–197.
10. Bitam, S., Zeadally, S., & Mellouk, A. (2016). Bio-inspired cybersecurity for wireless sensor networks. *IEEE Communications Magazine, 54*(6), 68–74.
11. Ahmad, M., Ikram, A. A., Leela, R., Masood, I., & Ulla, R. (2017). Honey bee algorithm-based efficient cluster formation and algorithm scheme in mobile ad hoc networks. *International Journal of Distributed Sensor Networks, 13*(6), 13–18.
12. Amouri, A., Alaparthy, V. T., & Morgera, S. D. Cross layer-based intrusion detection based on network behavior for IoT. In *Wireless and microwave technology conference.*

Chapter 8
Role of AI and Bio-Inspired Computing in Decision Making

Surekha Paneerselvam

Introduction

The basic theory of a human process is to establish an interaction with the outside world. This interaction involves decisions, which in turn lead to an argument of whether the decisions made are good or bad. Finally, people arrive at a conclusion, perceiving that they have made a good decision. But, in the process of decision making, they definitely seek help or support from someone to conclude whether the decision made is good. The supportive person tries to conceive the decisions and group them as structured, unstructured, and semi-structured. Structured decision problems lead to an optimal solution, for example, the decision made in optimizing the cost of travel between two cities. Unstructured decision problems do not have any predicted solution, for example, to make a choice between two equivalently good job offers or to make a decision to choose a life partner. Semi-structured decision problems definitely need a support system to develop alternate solutions and then pick the best out of them. Such support is provided with artificial intelligence (AI) techniques, and the unstructured decision-making process is simply referred as Intelligent Support Systems for Decision Making (ISSDM).

To a wider extent, AI mimics the process of human decision making, and this has also been proved by advanced AI researchers in several real-time situations. The human decision-making process is summarized by Pomerol and Adam [1], who claim that reasoning and recognition are the key parameters in decision making. They also claim that 'good' decisions are characterized by reasoning, which involves the process of weighing the alternatives and choosing the best decision. Most of the reasoning-based decision-making problems can be implemented using analytical techniques, and as a result, they can be implanted into ISSDM. On the other hand,

S. Paneerselvam (✉)
Amrita School of Engineering, Amrita Vishwa Vidyapeetham, Bengaluru, India
e-mail: p_surekha@blr.amrita.edu

© Springer Nature Switzerland AG 2020

G. R. Kanagachidambaresan et al. (eds.), *Internet of Things for Industry 4.0*,
EAI/Springer Innovations in Communication and Computing,
https://doi.org/10.1007/978-3-030-32530-5_8

recognition is the process which makes a decision without spotting out a proper reasoning. Such kind of recognition-based decision making has been implemented in emergency response situations such as fire services, pattern recognition applications, and authentication based on fingerprint and face, and in situations where immediate response is required. Decision making affects the neural system emotionally, and research has demonstrated the capability of modelling the emotional characteristics with decision making.

Decision support systems (DSS) were used since the early 1970s in order to assist several industrial activities in decision making. They were designed to provide an organized set of tools for the concerned decision-making situation and to improve the outcome for a chosen application [2]. A set of contradictory decisions are made for the problem under consideration, and the DSS chooses the best decision from the set of decisions.

A decision support system (DSS) is a system under the control of one or more decision-makers that assists in the activity of decision making by providing an organized set of tools intended to impart structure to portions of the decision-making situation and to improve the ultimate effectiveness of the decision outcome [2]. A decision support system (DSS) is a system which lets one or more people to make decision/s in a concrete domain, in order to manage it the best way, selecting at each time the best alternative among a set of alternatives, which generally are contradictory. Over the years, DSS evolved as a model-based set of procedures used to process the data for a given problem and to make suitable judgements. These model-based DSS were used to assist a human being in making decisions [3]. Later, the intellectual capabilities of human beings were combined with DSS using personal computers with a motive to improve the quality of the decisions made [4]. These computer-based support systems were later extended with a variety of possible decisions so that the best decision could be chosen for the problem under consideration. The DSS helped decision-makers to frame data and models based on an entity to solve structured as well as unstructured problems [5]. The basic feature of DSS lies in the model component for which quantitative techniques such as simulation, optimization, and statistics are used to represent a decision model [6, 7].

During the 1990s, DSSs were applied to perform tasks such as data analysis, monitoring and control, planning, prediction, recording data, and retrieval of data. To carry out these tasks, decision models such as optimization models based on linear programming, control algorithm models, statistical models based on regression, and simulation models were developed to assist humans in decision making.

The intelligent DSS were later evolved whose models were based on intelligent algorithms. These decision-making models were more commonly used due to their faster decision-making ability, consistency in the decisions made, and improved quality of decisions [8]. The computer-based decision-making systems incorporate an explicit decision procedure based on various theoretical principles to make decisions in an intelligent way [9]. The process of decision making is affected emotionally in the neural system either consciously and unconsciously. Research carried out during recent decades in Intelligent Support Systems for Decision Making (ISSDM) has shown the ability to handle emotion through decision making [10]. ISSDM use several artificial intelligence techniques to assist in the decision-making process.

These tools such as the artificial neural networks, fuzzy logic, and bio-inspired algorithms have been serving as an influential means to solve real-world problems that involve large amount of data. Some of the capabilities of these intelligent algorithms which have made them intelligent decision-makers are as follows:

- Understanding the situation and making sense out of the uncertainty or ambiguity
- Learn through experience
- React in a timely manner to a new situation (adaptive)
- Handling perplexing solutions
- Use knowledge to recognize various factors in a decision

ISSDM have been growing and emerging as useful tools for several applications by employing a variety of AI techniques. The basic concepts of AI techniques and their role in decision making are discussed in the following section.

ANN for Intelligent Decision Making

Artificial neural networks (ANN) are a powerful group of learning algorithms inspired from human nervous system. They are mainly used to approximate or estimate functions which have a huge dependency on the number of inputs they handle. ANN structure consists of interconnected neurons arranged in three layers, namely, the input, output, and hidden layer. The activation functions govern the manner in which the signals traverse between the layers through neurons. A weight is assigned to the connection made between neurons. Initially, the weights are assigned in random. The basic structure of a neural network is shown in Fig. 8.1. The input layer consists of neurons that process the inputs obtained from the real world and the output layer presents the results to the real world. The hidden layer lying between the input layer and the output layer consists of neurons that process information internally. The signals from the input layer are weighted and passed on to the hidden layer and in turn the weighted information is traversed to the output layer. This simple form of information processing network is referred to as feed-forward net work. Some networks send the output back to the hidden layer for further processing, known as the feedback network. Once the architecture of the neural network is formed, the network is prepared for the training and testing phases.

There are different ANN algorithms based on the learning concept, which are broadly classified into two categories, supervised learning and unsupervised learning. In supervised learning, the performance of the network is graded manually and the desired output is predicted, which in turn is presented to the network. In the unsupervised class of learning, the network is trained with updation of weights and tries to obtain the output based on the input information. This happens over a large number of iterations governed by the learning rules. The learning rules are administered by a control factor known as the learning rate. Slower learning rate results in a large convergence time, while faster learning rate results in inappropriate output. Hence a better choice has to be made as far as the learning rate is concerned.

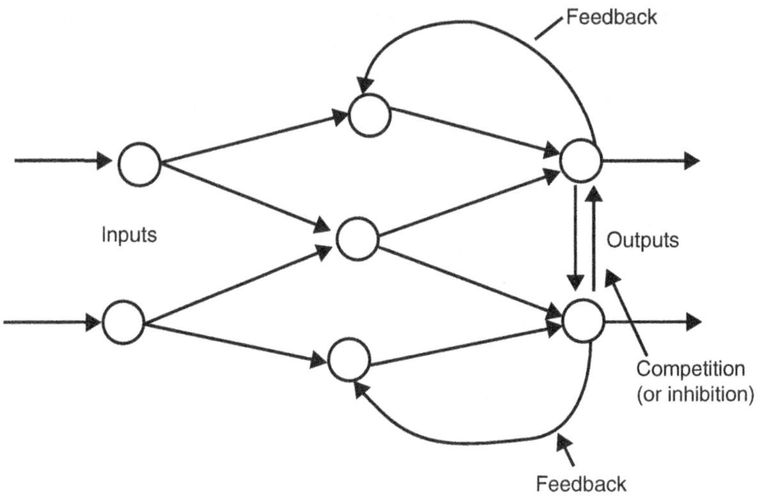

Fig. 8.1 Basic ANN model

The learning process is based on several learning laws, to name a few, Hopfield law, Hebb's rule, delta rule, extended delta rule, competitive learning rule, outstar learning rule, memory-based learning law, and Boltzmann learning law.

Major application areas of ANN are (a) prediction, (b) classification, (c) data association, and (d) data conceptualization. The perceptron network, back propagation network, and directed random search network belong to the group of prediction networks. Networks used for classification are learning vector quantization, counter-propagation network, and probabilistic neural networks, while Hopfield network, Boltzmann machine, Hamming network, and bi-directional associative memory are applied to data association–related problems. For data conceptualization, adaptive resonance network and self-organizing map have been more suitable networks.

In decision making, ANNs have been playing a vital role in making comprehensive decisions. This involves the need to make decisions through evolution and as a result consumes a lot of time. The evolution can be short term based on the course of an assignment, which has been influenced through feedback from the decisions obtained from the previous stages of the assignment. The evolution can also be long term, which evolves the transition over a course of life time. Whether the evolution is short term or long term, ANN should be capable of changing the current preference decisions based on the feedback obtained from the previous evolved decisions.

Fuzzy for Intelligent Decision Making

Fuzzy logic (FL) was proposed by Lotfi A. Zadeh in 1965 [11]. Fuzzy logic (FL) provides decision support based on the way humans reason them out. Fuzzy logic is used to represent uncertainty by assigning the input to a range of values between 0 and 1. The binary 0 and 1 refer to completely false and completely true, respectively. The range of values between the extremities is usually formulated using mathematic tools and later processed by computers, to achieve human-like thinking. The gap between qualitative and quantitative modelling can be bridged by applying FL. The input-output mapping of a FL model is quantitative, but internally the system is governed by a set of linguistic rules which are qualitative.

The fuzzy set is formed with elements in the universe, which contains all the possible elements for a related problem. The universe is defined by a membership function $\mu_A(x)$ where x is the number of elements in a base set. A variety of membership functions are defined for various applications, namely, triangular function, trapezoidal function, Γ-function, S-function, exponential function, and Gaussian function.

The process of obtaining the mapping between input and output using fuzzy logic is known as fuzzy inference. Decisions are made within the mapping based on rules. Fuzzy inference involves the processes namely fuzzification and defuzzification. The linking of input variables to output variables through expert knowledge and experience is based on the fuzzy rules. The rules fall under two basic categories such as Mamdani fuzzy rules and Takagi-Sugeno fuzzy rules. The inference system simply referred as a fuzzy expert system consists of modules such as fuzzifier, the inference engine, defuzzifier, and a fuzzy rule base.

The crisp inputs values are fuzzified and further sent to the rule base module. The rules are constructed using conditional IF-THEN statements which are used to control the output variable. The rules are applied to the fuzzy values, and the outputs are evaluated as fuzzy values. To send the fuzzy values to the real world, they are defuzzified and converted back to equivalent crisp values.

FL has been identified as suitable tools for decision making due to their capabilities such as:

- More choices to define uncertainty
- Consistency and redundancy of rule base and rules
- Addressing nonlinear issues also due to universal approximation capabilities
- Exact solutions can be obtained with rough approximation for a problem
- Simple models for a problem whose mathematical model is complex

Bio-Inspired Algorithms

Bio-inspired algorithms are a growing group of algorithms capable of searching the optimal solution from a search space at a quicker rate for a given optimization problem. Some of the existing conventional search algorithms follow a systematic procedure to obtain the optimal solution from the search space, and this takes a longer time for convergence. The disadvantages of conventional optimization algorithms are overcome by bio-inspired algorithms, in which biological behaviour of certain species are analysed to perform the optimal search. Hence the group of algorithms are also known as nature-inspired algorithms. These algorithms are framed based on the evolution resulting from interaction between species and interaction within the species in nature. The interaction between/within the species could be cooperative or competitive. Bio-inspired algorithms possess the following capabilities:

- Applicable to wide range of problems
- Few control parameters to tune the algorithm
- Better convergence rate while reaching the optimum value

Until recent years, ant colony optimization and particle swarm optimization inspired by colony of ants and swarm of bees were the basic bio-inspired algorithms applicable to solve constraint-based optimization problems. Later, behaviour of bacteria, fireflies, fruit flies, bats, and cuckoo birds were analysed to frame a wider class of bio-inspired algorithms. In this section, the variants of bio-inspired algorithms and their role in decision making is discussed.

Genetic Algorithms

Genetic algorithms (GA) were introduced as an optimization tool by John H. Holland in the early 1960s [12]. GA works by selecting individuals from a population of permissible solutions and then applying genetic operators, namely, crossover and mutation on them to choose the fitter individuals. The group of fitter individuals progresses to the next generation, and the evolution process continues iteratively until the convergence criteria is met. At the end of each iteration, the old population is rejected and the iteration continues using the new population. The basic pseudo-code of a GA is illustrated as:

```
Initialize the population of individuals
Evaluate the population through the fitness function
While no best solution do
    For every individual in the population
            Generate new population from old population
          Evaluate its fitness
          Selection
          Crossover
          Mutation
     End For
End While
```

Genetic algorithms have been used to solve a wide range of problems, namely, maximization/minimization problems, multi-objective optimization, and constraint-based optimization. Some of the industrial applications where GA has been successful are the job shop scheduling problem, vehicle routing problem, network routing, and load dispatch problem. The major challenges involved while solving problems using GA was in framing the fitness function, to understand the fitness of the solutions obtained through evolution, especially when the search space is large. This leads to scaling the solution space and thus leading to increase in the algorithmic complexity.

Thus GAs can be either applied to single or multi-objective problems whose fitness function can be evaluated with ease in a well-defined solution space. When the population size is increased for high dimension problems, then scaling is required as a result of which the algorithmic efficiency decreases.

The probable areas where GA can be applied to obtain convincing results are listed as follows

- Maximization/minimization problems
- Searching and sorting
- Job allocation (job shop scheduling)
- Parallel computation
- Routing, classification, coordination

Particle Swarm Optimization

Particle swarm optimization (PSO) developed by Dr. Eberhart and Dr. Kennedy in 1995 is a population-based stochastic optimization technique, inspired by collective social behaviour of fish schooling, bird flocking, or insect swarming [13]. A swarm of homogenous agents interact among themselves in the population, and they undergo the process of evolution to obtain the optimal solution. Unlike GA, PSO does not perform crossover and mutation during the evolution process. But when compared with GA, the merits of PSO are that the implementation process in PSO is simpler since there are a few parameters to adjust.

The PSO algorithm is initialized with a set of random particles which are the probable solutions to the defined problem. The algorithm then searches for the local optima by updating the successive generations. At the end of each iteration, every particle is updated based on two best values. One known as the pbest, this is the best solution that has been identified so far based on the fitness. The second value known as the gbest is the best value obtained so far in the population. PSO also defines another value known as the lbest, which refers to the best value in the population's topological neighbours. Upon finding the pbest and the gbest values, the particle updates its velocity and position using Eq. (8.1) and (8.2).

$$\mathrm{v}[\ ] = \mathrm{v}[\ ] + \mathrm{c1} * rand(\ ) * (\mathrm{pbest}[\ ] - \mathrm{present}[\ ]) + \mathrm{c2} * rand(\ ) * (\mathrm{gbest}[\ ] - \mathrm{present}[\ ]) \tag{8.1}$$

$$\mathrm{present}[\ ] = \mathrm{present}[\ ] + \mathrm{v}[\ ] \tag{8.2}$$

v[] is the particle velocity, present[] is the current particle (solution). pbest[] and gbest[] are defined as stated before. rand () is a random number between (0,1). c1, c2 are learning factors. Usually, c1 = c2 = 2.

The pseudo-code of the procedure is as follows:

```
Initialize swarm of particles (population)
While number of iterations is not maximum Do
    For each particle
        Calculate fitness value
        If the fitness value is better than the best fitness value
             (pBest) in history set current value as the new pBest
  End If
    Choose the particle with the best fitness value of all the
              particles as the gBest
              For each particle
        Calculate particle velocity according equation (a)
        Update particle position according equation (b)
End For
End While
```

To summarize, PSO is a population-based multidimensional multi-objective problem, with all the potential candidate solutions to the problem in a hyper-plane. When compared to GA, there is a possibility of highly scaling the problem space in case of complex problems.

The thrust areas where PSO is applied and proved to be performing better as an optimization solution are as follows:

- Job shop scheduling problems
- Adaptive learning problems
- Minimization/maximization problems
- Chaotic and oscillatory systems
- Searching and thresholding
- Location identification
- Resource management in a distributed environment

Ant Colony Optimization

Ant colony optimization (ACO) proposed by Dorigo in 1997 [14] is a popular optimization algorithm which was biologically inspired based on the behaviour of ants. ACO is a probabilistic technique which can be applied to real-world problems to find better paths through graphs. The search process is based on the trailing behaviour of ants between their colony and source of food thus establishing a path. The agents (artificial ants) in the algorithm locate the optimal solutions by moving through a population consisting of all possible solutions. During their movement, the agents record their positions and the quality of the food, analogous to the pheromone trailing in natural ants. This process of recording enables other agents in the population to identify better solutions.

The potential solutions are identified for a problem from a pool of possible solutions through probabilistic distribution. As the agents explore the possible solutions, each agent is updated locally. Further these possible solutions update the values of the trails such that the better quality solutions are selected in the subsequent evolutions. The process continues until an optimal solution is identified. It has been proved from several implementations that the algorithm completes the search process in a reasonable time.

The ACO algorithm works as follows:

```
Initialize the population with ants and pheromone values
While termination condition not reached do
  For every ant in the population
      Position ant on a starting node;
       For step size
             Obtain current state based on probabilistic transition rules
        End For
    Update the intensity of pheromone trail
    Compare and identify the best solution
   End For
    End While
```

The advantage of ACO lies in its convergence time mainly due to its property of inherent parallelism. ACO is suitable for solving dynamic optimization problems, continuous optimization problems, and NP-hard combinatorial problems. The property of distributed computation in ACO avoids premature convergence, hence the reason for faster convergence. The performance of the ACO algorithm reduces when the problem dimension increases.

ACO has been applied to a variety of real-world problems such as the following:

- Travelling salesman problem
- Network analysis and clustering
- Scheduling and routing
- Quadratic assignment problem
- Shortest path problem
- Economic dispatch problems
- Vehicle routing problem
- Gaming theory
- Graph colouring and set covering
- Forecasting
- Agent-based dynamic scheduling
- Parameter estimation

Artificial Bee Colony

The artificial bee colony (ABC) algorithm defined by Karaboga is a bio-inspired algorithm based on the foraging behaviour of honey bees [15]. Honey bees select site for hive, communicate, allot tasks among themselves, reproduce, forage, navigate, and lay pheromone. These capabilities of honey bees have been mimicked in the ABC algorithm. The colony of artificial bees is organized into three groups, namely, employed bees, onlookers, and scouts. The employed bees occupy one half of the colony, while the onlooker bees occupy the other half. One employed bee is allotted for each food source. In ABC algorithm, potential food sources are identified from the population initialized by the bees. Once the candidate solution (food source) is identified, then the validity (fitness) of the candidate solution is evaluated. In the successive stage, if a candidate solution of better fitness is discovered by the employed bees, then the existing candidate solution is discarded. Employed bees share all the information with the onlooker bees, and they try to improve the fitness of the discarded candidate solutions. If the onlooker bees are also not able to improve the fitness, then the candidate solutions are completely rejected from the population. The rejected solution is taken care of by the scout and replaces the population with another probable candidate solution. The process continues until the candidate solutions with better fitness are evolved. The pseudo-code of the ABC algorithm is as follows:

```
Initialize the population of candidate solutions
Evaluate the population of candidate solutions
While best solutions are not found so far do
    Produce new solutions for the employed bees to evaluate
    Apply selection based on Deb's method
    Calculate the probability values for the solutions
       Produce the new solutions for the onlookers from the
solutions
  Apply selection process based on Deb's method
     Determine the abandoned solution for the scout and replace
population
      Store the best solution achieved so far
End While
```

A wide range of studies have been carried out in the ABC algorithm over the years, based on their dance, communication, queen's behaviour, mating behaviour, navigation behaviour, and foraging behaviour. The ABC algorithms have been applied to constraint-based and unconstrained-based optimization problems in combination with differential evolution. The major application areas of ABC algorithm in the field of decision making are as follows:

- Numerical function optimization
- Network routing and allocation
- Benchmarking problems
- Searching
- Allocation and assignment
- Probability distribution

Fish Swarm Algorithm

Li et al. proposed the fish swarm algorithm (FSA) in 2002 [16]. The FSA is a population-based intelligent bio-inspired technique that simulates the schooling behaviour of fish under water. Usually, fish move close to each other in a group called swarm in order to search for food, protect themselves from predators, and in turn avoid collisions within the swarm. The behaviour in a swarm is characterized by leaping, chasing, swarming, and searching. While applying the swarm algorithm based on this behaviour to solve optimization problems, an individual fish within the swarm is considered as a point and is called the candidate solution. Collectively a population is formed which consists of several such candidate solutions or points. The environment in which the artificial fish moves forms the search space, in which the optimum solution is searched for.

Experiments have proved that the FSA has a capability of finding the global optimization solution without evaluating the local minimum. FSA has a strong ability to achieve global optimization by avoiding local minimum. The population is formed based on the position of the fish in a multidimensional search space. The behaviour of the fish during the food search is weighted by a term known as the food satisfaction factor. The distance between two fish in the search space is characterized by the Euclidean distance. The fish in the swarm searches for the food in a random manner in the search space moving in a direction based on high food concentration. The objective of the optimization technique is to minimize the food satisfaction factor. Within the population, fish establishes a swarming behaviour to satisfy the food intake and during the process, they tend to attract new swarm members.

Once a fish identifies a food location, the individuals in its neighbourhood follow. Within the visual region of the fish, some fish in the population will be able to find more amount of food when compared to the others. These fish will obviously follow the best one to increase the food satisfaction percentage. The basic parameters involved in FSA are the visual distance, maximum step length, and a crowd factor. The algorithm efficiency is mostly dominated by the two parameters, namely, visual distance and maximum step length.

The pseudo-code of the fish swarm algorithm is as follows:

```
Initialize the swarm of fish in a random manner.
While stop condition is reached do
 for each fish
  Evaluate the fitness
 do step Follow (based on the food search behaviour)
 if (Follow has failed) then do step Swarm
     if (Swarm has failed) then do step Prey
     end
  end
  end for
end while
```

To summarize, FSA is a population-based multi-objective problem, with the ability to solve complex nonlinear high-dimensional problems and with all the potential candidate solutions to the problem in a hyper-plane. The algorithm is capable of converging faster since it requires very few parameters to be tuned for optimization. Moreover, FSA does not possess the crossover and mutation processes used in GA, so the algorithm converges faster.

The common areas where FSA is applied and proved to be performing better as an optimization solution are as follows:

- Function optimization
- Parameter estimation
- Combinatorial optimization
- Least squares support vector machine
- Geotechnical engineering problems

Firefly Algorithm

Firefly algorithm (FA) proposed by Yang [17] was inspired by the flashing behaviour of fireflies and were used to address the non-convex objective functions in NP-hard problems. These problems have equality- and inequality-based constraints, and the firefly algorithm has been proved by providing suitable optimal solutions to these problems. The algorithm employs a population-based search with agents that are analogous to fireflies. The algorithm uses a learning mechanism that facilitates good tuning of parameters thus balancing the exploration and exploitation. The algorithm iterates with numerous agents that communicate with each other. The bioluminescent glowing lights are used as a media through which the information is shared. These glowing lights enable the fireflies to explore the search space more effectively when compared to the distributed random search. Each fly is attracted by the brighter firefly within its neighbourhood. The optimization algorithm has the following rules based on the biological behaviour:

- The fireflies are attracted towards brighter ones irrespective of their sex.
- The degree of attractiveness of a firefly and brightness of a firefly are proportional to each other.
- The brightness or light intensity of a firefly is determined by the value of the objective function of a given problem.

The pseudo-code of the firefly algorithm is as follows:

```
Initialize population of fireflies
Determine the Light Intensity based on the fitness function
Define a light absorption coefficient
While (maximum numbers of Generations have not reached) do
For each firefly
      Movement of flies towards attractive flies
              Attractiveness varies with distance according
to the light absorption coefficient
       Evaluate new solutions and update light intensity
End For
Determine the current best based on rank
End While
```

The behaviour based on attraction towards brightness and the ability of dealing with multimodality has made the firefly algorithm to deal with multimodal functions to solve NP-hard optimization problems.

Some of the potential areas in which the firefly algorithm has been applied successfully are as follows:

- Multi-objective optimization
- Dispatch problems
- Digital image compression
- Feature selection
- Chaotic problems
- Nonlinear problems – structural optimization
- Multimodal design problems
- Forecasting problems – stock exchange
- Antenna design optimization
- Nonlinear optimization
- Load dispatch problems
- NP-hard scheduling problems
- Dynamic environment problems
- Classifications and clustering
- Training ANN

Bacterial Foraging Algorithm

Bacteria foraging optimization (BFO) algorithm proposed by Passino in 2002 [18] is a stochastic global search technique inspired biologically by the foraging behaviour of *E. coli*. The algorithm is based on the biological behaviour in which the organisms with less capabilities are eliminated and the others survive through the process of natural selection. The process involves three stages, namely, locating the food, handling the population, and consuming the food. Such foraging strategy happens at an individual level or at a group level. During the food search (evolutionary) process, the bacteria try to maximize their energy consumption per unit time spent in search of food. All bacteria in the population establish communication among themselves through signals. The foraging decisions are based on the energy consumption and the communication through signals. The agents are the bacterium, and they are governed by two actions: tumbling and swimming.

The tumble action involves the movement of the agents in a random direction in search of an optimal solution, while in the swimming action the bacterium move around in the population by taking small steps in search of food. The process continues, and as a result of reproduction during the chemotactic movement, the best agents are retained in the population, while the inefficient ones are eliminated. The best agents in the population are those which have better objective function and maximum energy per unit time. The algorithm uses an elimination-dispersal operator which comprehends the environmental changes due to the elimination of agents.

Though the algorithm is simple and easy to implement, it suffers from poor convergence rates. The working of the BFO algorithm is as follows:

```
    Initialize population
    Evaluate the fitness
    While number of iterations not reached do
          For every chemotactic process
                 Evaluate the fitness function
                 Perform tumbling and swimming
          End For
          While chemotaxis operation do
                 Evolution through Reproduction
          End While
                  Elimination Dispersal operator to retain the best
individuals
          End For
    End While
```

BFO has been found application to a variety of real-world problems due to its excellent global searching capability and proved its effectiveness over variants of GA and PSO. On the other hand, BFO has also failed with poor convergence criteria. The algorithm is also sensitive to the step size value required for the tumbling, and hence it is difficult to generalize an explicit constraint-handling mechanism for the algorithm. Certain improvements in the mathematical modelling and adaptation strategies of the BFO algorithm might make the algorithm more efficient.

Some of the application areas where BFO has been successful are as follows:

- Multi-objective function optimization
- Load forecasting and compensation
- Harmonic analysis in power systems
- Optimal tuning of PID controller
- Load dispatch and unit commitment
- Machine learning
- Maximization and minimization
- Job shop scheduling

Cuckoo Search Optimization

The cuckoo search optimization (CSO) algorithm for single and multi-objective nonlinear optimization problems was proposed by Yang and Deb in 2002. The algorithm draws its inspiration from the biological breeding behaviour of cuckoo birds which they lay their eggs in the nest of host birds. They remove the eggs of the host bird and try to increase the hatching probability of the cuckoo eggs. The individuals for evolution in this search algorithm are generated through Levy flight mechanism, which is a special class of random walk with irregular step lengths based on probability distribution. There are three types of brood parasitism adopted by the cuckoos namely the intraspecific, cooperative, and nest takeover mechanisms. In intraspecific brood parasitism, the cuckoos lay eggs in the host bird's (same species) nest and later does not care for the eggs [19]. Cooperative breeding refers to the pairing of two or more females with the same male, and they lay their eggs in the same nest in a cooperative manner and are mutually cooperative in providing parental care [20]. The nest takeover [21] breed mechanism refers to a cuckoo occupying another host birds' nest to lay its eggs.

The cuckoo search algorithm is based on three basic rules:

(i) Each cuckoo lays an egg one at a time and puts it in nest chosen at random.
(ii) The nests with good quality will progress to the next generation.
(iii) The host bird in the host nest is capable of discovering a foreign egg based on probability.

The Levy flight mechanism provides the random walk in which the step length is obtained based on Levy distribution. The new solutions are generated through the random walk process. The movement is always around the best solution obtained so far; hence, the local search is always faster. Some fraction of new solutions are generated by far-field randomization, so that they are lying far from the current best solution. This enables the algorithm from avoiding getting trapped in the local optimum.

Based on the biological behaviour of the cuckoo birds and the above-mentioned rules, the algorithm for finding an optimal solution is framed as follows:

```
Initialize a population of host nests
While (stopping condition not reached) do
     Evaluate the fitness of a randomly chosen cuckoo
     Select a nest from the population
     if(fitness of cuckoo is higher than the fitness of the nest)
     {
             Replace the fitness of the nest with the fitness of
the cuckoo
     }
      Discard the worse nests and build new ones
     Retain the best nests (solutions);
     Rank the nests
     Obtain the current best
End While
```

The application of Levy flights in CSO has proved to be an efficient approach to solve optimization problems. The random walk property of Levy flight mechanism helps the CSO algorithm to converge much faster, thus improving the algorithmic efficiency. In order to obtain the optimal solution, a large number of iterations are required, which is one of the drawbacks of the CSO algorithm. This increased amount of iterations is obtained when the discovery rate is small and the Levy step size is high. On the other hand, when the discovery rate is high and the Levy step size is small, the algorithm converges faster, but the best solution may not be obtained. Hence, a trade-off has to be maintained between the discovery rate and the Levy step size.

The CSO algorithm has been successful in the following problem areas:

- Optimization of ANN parameters
- Gradient-based optimization
- Multi-objective scheduling and allocation
- Phase equilibrium problems
- Optimization in cluster centres
- Reliability optimization
- Economic load dispatch
- Path identification for network analysis

Fruit Fly Algorithm

Fruit fly optimization algorithm (FFA) is an evolutionary algorithm introduced by Pan [22]. The algorithm is inspired by the biological behaviour of fruit flies. The fruit flies have excellent smell and sight senses, using which they find their food.

It has been proved in biological sciences that these group of flies can smell food even at a distance of 40 km. The foraging behaviour in search of food is performed in two stages. During the first stage, the flies identify the food through their smell (osphresis organ), and in the second stage they get near the food through their sensitive vision. Usually, the smell phase is shorter than the vision phase.

The procedure of the FFA is summarized as follows:

```
Initialize the fruit fly swarm in a random manner
  While (maximum number of iterations not reached) do
  Each individual randomly searches the food source and distance
is obtained
  Smell concentration value is calculated for every fruit fly
  Identify the location with best concentration value
  Position the swarm for successive iterations
  End While
```

The fruit fly algorithm is capable of finding the global optimum at a faster rate with improved accuracy due to the smell concentration parameter. Due to this parameter, the algorithm converges without falling into the local minima, hence improving the robustness. The update strategy is very simple in the fruit fly algorithm, but it is difficult to define a standard update strategy during the initialization process.

The algorithm is still growing and finding its roots in several applications. Some of the areas where the algorithm has found scope are as follows:

- Military applications
- Medicine
- Management and finance
- In combination with data mining techniques

Bat Algorithm

The bat algorithm (BA) was introduced by X.S Yang in 2010 [23] for continuous problem domain. Later a binary version of the bat algorithm was introduced by Mirjalili et al., in 2014 for discrete problem domain. The algorithm evolves on the basis of echolocation behaviour of bats. Bats use echoes of sounds (echolocation) emitted by them and then they navigate through their surroundings. Bats use this approach to find their prey during nights. Using the echolocation process as the basis, the bat algorithm was proposed to obtain the optimal solution for a multi-objective optimization problem. Bats have the capability to adjust their flight velocity and the frequency of the sound (pulse emission) while searching for food. These parameters, namely, the flight velocity and the pulse emission are tuned adaptively so that their intensity decreases once the bat has identified a potential prey.

The algorithm is governed by three important rules.

1. The distance between the bat and the food is governed by the echolocation capability of bats. This is in turn used for sensing barriers in the darkness.
2. The flight movement of bats during food search is random based on its characteristic features such as velocity, frequency, and loudness.
3. Bats also have the capability to change the frequency of the sound (loudness).

The algorithm evolves around a population of bats. An initial position is assigned to each bat in random which forms the initial population. The bats move within the population and try to obtain the local optimum solution and from where it tries to obtain the global optimum solution. During the process, the parameters, namely, pulse emission and loudness are updated. The process continues until the best solution is obtained.

The procedure of bat algorithm (BA) is shown as follows:

```
Initialize population of the bats
Initialize parameters with random values
While (Stopping criterion not met) do
    Generate new solutions
    Compare pulse rate of current with the random one
    Select solution among the best solution
          Generate a new solution based on random flight
          Compare loudness, pulse frequency of current with the
random
                Accept the new solutions
          Update pulse rate and reduce loudness
End while
```

The bat algorithm has been found suitable for nonlinear and multimodal problems. It is mostly suitable for high-dimensional problems where convergence criteria is a major challenge.

Bat algorithm has been successful in the following applications:

- Structural design optimization
- Unit commitment and economic load dispatch
- Adaptive learning problems
- Path planning and scheduling
- Network routing and analysis
- Constraint-based optimization
- Multi-objective optimization

Role of ISSDM in Industry 4.0

During the recent years, the word 'Industry 4.0' has become so common and is trying to establish its roots in the future of industrial production. The initiative was started by the Germans, and then the concept has extended universally to understand and sort out means for the forthcoming industrial revolution [24]. In a smart industry environment, almost all the processes are integrated. Raw material, machines, products, human beings, logistics, communication, and IT-related operations are all interlinked with each other with the objective to improve the overall production. The intelligent support systems for decision making are modelled to increase the mass production in smart industries by decentralizing the control operations and decisions. This facilitates amendments in the production process thus increasing the demand for mass customization. Most of the industries have already established the Industry 4.0 drive and have been successful in the market. On the other hand, some of the industries are still looking for prospects and advantages to join the Industry 4.0 drive. In spite of the various dilemmas, companies are trying to get into the Industry 4.0 drive due to the increase in competitiveness, fault diagnosis, fault reduction, need to use currently growing technologies, and skilled manpower.

In Industry 4.0, decisions have to be taken such that the efficiency of the manufacturing process is maintained throughout the several phases of operation. There is a large amount of data to be handled which has paved the way for the need for suitable decision models. To enhance the error analysis and prediction of a particular situation to take decisions, intelligent tools such as AI and bio-inspired algorithms are used in decision support systems, thus improving the automation process in an industry. In Industry 4.0 paradigm, the decision-making models are decentralized, in which the production entities are broken down into numerous sub-entities. Each sub-entity works individually trying to make its decisions for its problem definition. All the sub-entities are collaborated and through certain mechanisms which co-ordinate and exchange information between themselves. To handle such complex tasks, ISSDM is applied so that all the entities work in an autonomous fashion based on the knowledge of the models and knowledge of human decisions.

In any industrial process, there are a wide range of parameters that have to be tuned and controlled very often. This is a challenging task for the AI or bio-inspired algorithm chosen for the purpose, since the major drawback of these algorithms lies in tuning of its parameters. Therefore depending upon the industrial process, the intelligent algorithm is chosen wisely. The general algorithm to apply ANN and bio-inspired algorithm to an Industry 4.0 application is as follows:

```
Step 1: Define the objective function and a set of process variables for the problem under consideration
Step 2: Train the ANN to simulate the relation between the process variables and the objective function.
Step 3: Define the constraints related to the process variable and the convergence criteria
Step 4: Select a new set of process variables based on the optimization technique
Step 5: Apply ANN to simulate the process behaviour and the objective function
Step 6: Check for stopping condition otherwise repeat Steps 3 to 5
```

The above algorithm is a hybrid combination of ANN and bio-inspired algorithm. The need for hybridization is to reduce the time taken to tune the algorithmic parameters and improve the convergence rate.

Conclusion

Recent advances in high computing power have led to handling large amount of data and information. As a result, the automation in several industrial applications have become more complex especially in the areas of finance, and health care. Thus AI-based tools for decision making has becoming an emerging research during the recent years. Most of the real-world problems related to the Industry 4.0 paradigm are complex, uncertain, and dynamic. AI and bio-inspired algorithm or a hybrid combination of these algorithms has proved their capability in solving such problems. They have also proved to handle situations in changing environments, whenever a real-time decision is required.

Humans are excellent decision-makers – they take decisions based on past experience, skill set, and knowledge. Human-machine interaction is again evolving to train the machines to take decisions based on human experience. Such communication between man and machine is based on autonomy, interaction, and personification. Such intelligent interfaces have been emerging with the aid of AI and bio-inspired algorithms. Manyika et al. [25] stated that 'sophisticated analytics can substantially improve decision making'. Using ISSDM, improvement can be done in decision making by collecting relevant information, providing specific information for better performance, conducting experiments to validate the performance, forecasting, and casting the present scenario.

Several challenges are still remaining in intelligent decision making for future researchers to study and propose better solutions. There are a wide range of unsolved complex real-world problems due to the following characteristics:

- Difficult for humans to understand and obtain the relation between the operational variables

- Variables are dynamic in nature
- Involve events that are difficult to understand and perceive
- Involve large amount of data

The challenge for ISSDM is to provide solution for such kind of problems leading a new path for innovation thus providing efficient and effective decisions.

References

1. Pomerol, J.-C., & Adam, F. (2008). Understanding human decision making – A fundamental step towards effective intelligent decision support. In *Intelligent decision making: An AI-based approach* (pp. 3–40). Berlin, Heidelberg: Springer.
2. Marakas, G. M. (1999). *Decision support systems in 21st century – US edition*. Upper Saddle River, London: Prentice Hall.
3. Little, J. D. C. (1970). Models and managers: The concept of a decision calculus. *Management Science, 16*(8), B-466.
4. Keen, P. G. W., & Scott Morton, M. S. (1978). *Decision support systems: An organizational view*. Boston: Addison Wesley.
5. Beulens, A. J. M., & Van Nunen, J. A. E. E. (1988). The use of expert system technology in DSS. *Decision Support Systems, 4*(4), 421–431.
6. Castle, E. N., Becker, M. H., & Smith, F. J. (1972). Farm business management: the decision making process. *Farm business management: the decision making process*. 2nd edn.
7. Bell, J. (1992). Discretionary decision-making: A jurisprudential view. *The uses of discretion*, 89–111.
8. Haagsma, I. G., & Johanns, R. D. (1994). *Decision support systems: An integrated and distributed approach* (Vol. 01821, pp. 205–212). Billerica: Computational Mechanics, Inc..
9. Fox, J., & Das, S. (2000). *Safe and sound: Artificial intelligence in hazardous applications* (p. 307). Menlo Park: AAAI Press/MIT Press.
10. Santos, M. F., Portela, F., & Vilas-Boas, M (2011). INTCARE: Multi-agent approach for real-time intelligent decision support in intensive medicine. (pp. 364–369).
11. Zadeh, L. A. (1965). Information and control. *Fuzzy Sets, 8*(3), 338–353.
12. Goldberg, D. E., & Holland, J. H. (1988). Genetic algorithms and machine learning. *Machine Learning, 3*(2), 95–99.
13. Eberhart, R., & Kennedy, J. (1995). A new optimizer using particle swarm theory. In *MHS'95. Proceedings of the Sixth International Symposium on Micro Machine and Human Science*. IEEE.
14. Dorigo, M., & Gambardella, L. M. (1997). Ant colony system: A cooperative learning approach to the traveling salesman problem. *IEEE Transactions on Evolutionary Computation, 1*(1), 53–66.
15. Karaboga, D., & Basturk, B. (2007). A powerful and efficient algorithm for numerical function optimization: Artificial bee colony (ABC) algorithm. *Journal of Global Optimization, 39*(3), 459–471.
16. Li, X.-l. (2002). An optimizing method based on autonomous animats: Fish-swarm algorithm. *Systems Engineering-Theory & Practice, 22*(11), 32–38.
17. Yang, X.-S. (2010). Firefly algorithm, stochastic test functions and design optimisation. *International Journal of Bioinspired Computation, 2*(2), 78–84.
18. Passino, K. M. (2002). Biomimicry of bacterial foraging for distributed optimization and control. *IEEE Control Systems Magazine, 22*(3), 52–67.
19. Ruxton, G. D., & Broom, M. (2002). Intraspecific brood parasitism can increase the number of eggs that an individual lays in its own nest. *Proceedings of the Royal Society B: Biological Sciences, 269*(1504), 1989–1992.

20. Gibbons, D. W. (1986). Brood parasitism and cooperative nesting in the moorhen, Gallinula chloropus. *Behavioral Ecology and Sociobiology, 19*, 221–232.
21. Payne, R. B., Sorenson, M. D., & Klitz, K. (2005). The Cuckoos, Oxford University Press.
22. Pan, W.-T. (2012). A new fruit fly optimization algorithm: Taking the financial distress model as an example. *Knowledge-Based Systems, 26*, 69–74.
23. Yang, X.-S. (2010). A new metaheuristic bat-inspired algorithm. In *Nature inspired cooperative strategies for optimization (NICSO 2010)* (pp. 65–74). Berlin, Heidelberg: Springer.
24. Schlaepfer, R. C., Koch, M., & Merkofer, P. (2015). Challenges and solutions for the digital transformation and use of exponential technologies. *Deloitte*. http://deloitte.com/content/dam/Deloitte/ch/Documents/manufacturing/ch-en/manufacturing-industry-4-0-24102014.pdf
25. Brown, B., Chui, M., & Manyika, J. (2011). Are you ready for the era of 'big data'. *McKinsey Quarterly, 4*(1), 24–35.

Chapter 9
Smart Recognition System for Business Predictions (You Only Look Once – V3) Unified, Real-Time Object Detection

Allumallu Veera Venkata Susmitha

Introduction

Object recognition is viewed as a standout amongst the most difficult issues in this field of PC vision, as it includes the mix of item arrangement and object confinement inside a scene. As of late, Deep Neural Network systems (DNNs) have been shown to accomplish better item discovery execution through different methodologies, with YOLOv2 (an enhanced You Only Look Once model) being one of the cutting edges in DNN-based object location strategies in both speed and precision. Even though YOLOv2 can accomplish constant execution on a ground-breaking GPU, it stays exceptionally trying to utilize this methodology for continuous item identification in video on installed figuring gadgets with constrained computational power and restricted memory. In this project, we propose another system called Fast YOLO V3, a quick You Only Look Once structure that quickens YOLOv2 to have the capacity to perform object location in video on inserted gadgets in a continuous way. To begin with, we use the developmental profound knowledge system to advance the YOLOv3 arrange engineering and deliver an improved design that has 2.8× fewer parameters with only an ~2% IOU drop. To additionally lessen control utilization on installed gadgets while looking after execution, a movement versatile surmising technique is brought into the proposed Fast YOLO structure to decrease the recurrence of profound derivation with O-YOLOv3 dependent on transient movement qualities.

A. V. V. Susmitha (✉)
National Chiao Tung University, Hsinchu, Taiwan

© Springer Nature Switzerland AG 2020
G. R. Kanagachidambaresan et al. (eds.), *Internet of Things for Industry 4.0*,
EAI/Springer Innovations in Communication and Computing,
https://doi.org/10.1007/978-3-030-32530-5_9

Literature Survey

We have investigated a couple of updates to YOLO. We made a heap of little arrangement changes to enhance it. It is more noteworthy than last time, yet dynamically correct. It is still snappy, notwithstanding. We present YOLO v3 as another way to deal with object discovery. Earlier work on item location repurposes classifiers to perform recognition, while we outline object location as a relapse issue to spatially isolated bouncing boxes and related class probabilities. A solitary neural system predicts jumping boxes and class probabilities straightforwardly from full pictures in a single assessment. Since the entire recognition pipeline is a solitary system, it very well may be improved start to finish on identification execution.

We brought together engineering, and it is amazingly quick. Contrasted with best in class location frameworks, YOLOv3 makes more confinement blunders yet is far less inclined to anticipate false identifications where nothing exists. At long last, YOLOv3 adapts extremely broad portrayals of items. It beats all other location techniques, including DPM and R-CNN, by a wide edge while summing up from normal pictures to works of art on both the Picasso Dataset and the People-Art Dataset.

Motivation Behind Using YOLOv3

Why Yolov3?

Among cutting edge techniques for profound learning object location (Faster R-CNN, SSD, YOLO, etc.), Yolov3 emerges because of its incredible harmony among speed and exactness. In Yolov3, each info picture is partitioned into an S × S lattice. For every cell in the network, some jumping boxes forecasts are created at the same time with class probabilities/scores for foreseeing objects related with that lattice cell. Each score reflects how sure the model is that the container contains a specific class of item [1–5].

Comparison of YOLOv1, v2 & v3

In Fig. 9.1, we refer to the various parameters and their performances and took this figure from the paper referred to in reference [6], and now we will see the comparison of YOLOv3 and why exactly we are using YOLOv3. One advantage of picking this characterization is that when YOLO cannot recognize the sort of plane, it gives a high score to the plane as opposed to compelling it into one of the sub-classes. At the point when YOLOv3 sees the farther picture, it just back propagates arrangement misfortune to prepare the classifier. YOLOv3 finds the bouncing box that predicts the most elevated likelihood for that class, and it figures the arrangement misfortune just as those from the guardians [7].

	YOLO								YOLOv2
batch norm?		✓	✓	✓	✓	✓	✓	✓	✓
hi-res classifier?			✓	✓	✓	✓	✓	✓	✓
convolutional?				✓	✓	✓	✓	✓	✓
anchor boxes?				✓	✓				
new network?					✓	✓	✓	✓	✓
dimension priors?						✓	✓	✓	✓
location prediction?						✓	✓	✓	✓
passthrough?							✓	✓	✓
multi-scale?								✓	✓
hi-res detector?									✓
VOC2007 mAP	63.4	65.8	69.5	69.2	69.6	74.4	75.4	76.8	**78.6**

Fig. 9.1 Comparison of YOLOv1 & YOLOv2

Design of the Model

Bounding Box Prediction

Following YOLO9000, our framework predicts jumping boxes utilizing measurement groups as stay boxes. The below figure is from the paper cited in reference [8]. Then it predicts four organizations for each bouncing box, t_x, t_y, t_w, t_h, on the off chance that the cell is counterbalanced from the upper left corner of the picture by (c_x, c_y) and the bouncing box earlier has width and tallness p_w, p_h (Fig. 9.2).

Class Prediction

Every cell predicts the classes; the ricocheting box may be contained using multi-label portrayal. We do not use a softmax as we have found it is futile for good execution, rather we simply use self-ruling determined classifiers. During set up, we use betray entropy disaster for the class desires. This arrangement causes movement to progressively amazing territories such as the Open Images Dataset [9]. In this dataset there are many covering marks. Using a softmax powers the assumption that every cell has exactly one class, which is routinely not the circumstance. A multi-label approach is a superior model for getting precise data [10–14].

Feature Extraction

In our task, we have used another framework for performing Feature Extraction. Our new framework is a blend approach between the framework used in YOLOv2, Darknet-19, and that advanced waiting to orchestrate stuff. Our framework uses

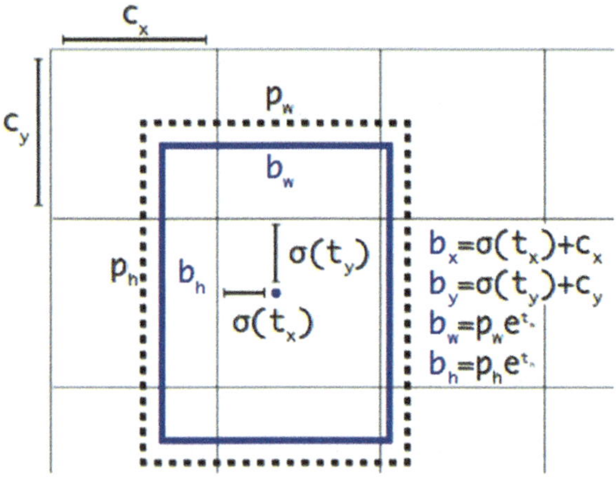

Fig. 9.2 Bounding box prediction. b_x, b_y, b_w, $b_h = x$, y center coordinates, width and height of our prediction. t_x, t_y, t_w, t_h = Network outputs. c_x, c_y = Top left coordinates of the grid. p_w, p_h = Anchor dimensions for the box

dynamic 3 × 3, and 1 × 1 convolutional layers now have some backup course of action affiliations as well and are on a very basic level greater. It has 53 convolutional layers, so we call it Darknet-53. Every framework is set up with indistinct settings and attempted at 256 × 256 single item precision. Thus, Darknet-53 performs on standard with best in class classifiers with less coasting point assignments and more speed. Darknet-53 is better than ResNet-101 and 1.5× speedier. Darknet-53 has equivalent execution to ResNet-152 and is 2× faster. Darknet-53, moreover, achieves the most amazing evaluated skimming point exercises each second. This infers the framework structure better uses the GPU, making it progressively able to evaluate and thus snappier. That is for the most part because ResNets currently have such numerous layers and are not very profitable [15–22].

Training

We train on full pictures with no hard-negative mining or any of that stuff. We use multi-scale preparing, bunches of information growth, clump standardization, all the standard things. We utilize the Darknet neural system structure for preparing training and testing.

- End-to-End Convo-Net
- Input = Images from coco dataset (common objects in context)
- Metafile: Indicating the coordinates and the class of each of the objects in image
- Size of input = 416 × 416
- Minimum confidence threshold = 0.5
- Non-max suppression threshold = 0.3
- Load pre-trained Darknet-53 and add 53 detection layers to it
- Output of Network: Position of bounding box (x, y, w, h) class of object, confidence

Network Architecture

The above Fig. 9.3 demonstrates the system design of YOLOv3 alluded to in [8]. YOLO might be a convolutional system, and its definitive yield is created by applying a 1 × 1 portion on a component outline. In YOLOv3, the recognition is done by applying 1 × 1 discovery pieces on highlight maps of three totally extraordinary sizes at three better places inside the system. The state of the location bit is 1 × 1× (B × (5 + C)). Here B is that the assortment of bouncing boxes a cell on the component guide will foresee, "5" is for the four jumping box traits and one item certainty, and C is that the assortment of classifications. In YOLOv3 prepared on coco palm, B = 3 and C = 80, in this manner the portion estimate is one × one × 255. The element outlined by this portion has indistinguishable tallness and measurement of the past component delineate has location properties on the profundity as portrayed higher than.

YOLOv3 makes forecasts at three scales that square measure precisely by down examining the size of the information picture by 32, 16, and 8, severally. The primary recognition is made by the 82nd layer. For the essential 81 layers, the picture is down inspected by the system, such that the 81st layer includes a walk of 32. If we have an image of 416 × 416, the resultant component guide would be of size 13 × 13.

At that point, the element outline layer 79 is exposed to numerous convolutional layers before being up inspected by 2× to measurements of 26 × 26. This element outline is then profundity connected with the element delineate layer 61. A comparable method is pursued yet again, wherever the component outline layer 91 is exposed to a few convolutional layers before being profundity linked with an element delineate layer 36. Here similar examination objects might be picked inside a similar article by various layers.

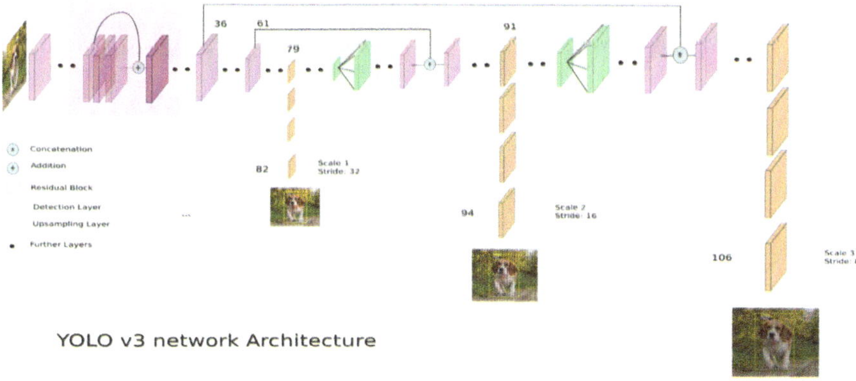

Fig. 9.3 YOLOv3 network architecture

Results

Real Time Image Testing (Fig. 9.4)

Input: Real image
Output: Position of bounding boxes (*X*, *Y*, *W*, *H*) Score, Class ID, Confidence

Art Work Image Testing (Fig. 9.5)

Input: Art work image
Output: Position of bounding boxes (*X*, *Y*, *W*, *H*) Score, Class ID, Confidence

Sample Video Testing (Fig. 9.6)

Input: Sample video
Output: Position of bounding boxes (*X*, *Y*, *W*, *H*) Score, Class ID, Confidence

Fig. 9.4 Real time image testing

Fig. 9.5 Art work image testing

Fig. 9.6 Sample video testing

Summary

From the above results, we can say that we have a new approach to identify objects with this advanced technology. Still, there are a few constraints in the performance of YOLO as the model has to be detected as well as classified by using only Prime neural network. The optimization loss can be trained on the end-to-end network for better detection performance to avoid the localization errors. The limitation occurs when the Faster RCNN can perform with fewer background errors, whereas our model faces some struggles such as more background errors. Unlike other models, our model collects the data from bounding boxes, which generalizes the objects with improper configurations. Some of the main constraints to work on our model are as follows:

1. *Batch Normalization*: In our model Batch Normalization has helped in regularization, which resulted in suppressing the Overfitting cause.
2. *High Resolution Classifier*: Generally, the images that we have used in our model are 256 × 256. Thus, we can say that it increased the input size while the model I trained on the DarkNet ImageNet database. There is a 4% increment on MAP.
3. *Anchor Boxes*: There is a tremendous change in the accuracy by utilizing the anchor boxes, whereas in YOLOv2, they are not used, but are replaced by the means of input dimensions using K-means clustering to define the anchor boxes.
4. *Multi-Scale Training*: In the perspective of using different image sizes for the purpose of detecting objects, the performance was not up to the level.
5. *Fine-Grain features*: The prediction detection in model is on 13 × 13 feature map, which is much smaller than other YOLO models. This feature helps us in localizing both larger and smaller objects efficiently.
6. *DarkNet*: The Architecture we used in our model is better compared with YOLOv2. It has 19 convolutional layers, five Maxpooling layers, and a softmax layer on the top of the convolutional layer for the purpose of classification.

Let us present a short conclusion of Improvements in YOLOv3:

Bounding Box Predictions YOLOv3 simply like YOLOv2 utilizes dimension clusters to produce Anchor Boxes. Presently, as YOLOv3 is a solitary system, the misfortune for objectiveness and classification should be determined independently, yet from a similar system. YOLOv3 predicts the objectiveness score utilizing logistic regression where 1 means total cover of bounding box earlier over the ground truth object. It will anticipate just 1 bounding box earlier for one ground truth object and any mistake in this would lead to both classifications just as detection misfortune. There would likewise be other bounding box priors that would have an objectiveness score more than the edge; however, not exactly the best one, for these mistakes will develop for the detection loss and not for the classification loss.

Class Predictions YOLOv3 utilizes free logistic classifiers for each class rather than an ordinary softmax layer. This is done to make the multi-label classification. Take a model where a lady appears in the image and the model is prepared on both individual and lady, having a softmax here will prompt the class probabilities that have been partitioned between these two classes with state 0.4 and 0.45 probabilities. Be that as it may, autonomous classifiers tackle this issue and give a yes versus no likelihood for each class, similar to what is the likelihood that there is a lady in the image would give 0.8 and what is the likelihood that there is an individual in the image would give 0.9, and we can name the item as both individual and lady.

Predictions over scales To help detect shifting scales, YOLOv3 predicts boxes at three distinct scales. At that point, highlights are separated from each scale by utilizing a technique like that of feature pyramid systems.

Applications

Detection of objects is the primary task in Computer Vision

1. For detection of objects in real time, YOLO is a brilliant neural network.
2. Uses Non-Maximal Suppression for the purpose of best bounding box.
3. For the use of standard layers as convolutional with 3×3 kernel, Max-pooling with 2×2 kernel.

Future Scope

What can be improved for the upcoming Model?

1. For smaller and medium objects, the average precision should be improved, so it can perform better than Faster RCNN.
2. To decrease the localization errors, MAP must be increased efficiently.
3. Currently, for DarkNet implementation, we used C language, but Python implementation can be more efficient.

Because we know that YOLOv3 is better—stronger but not faster— future work could be more concentrated on the perspective of making it faster. As we need the most accurate algorithm, the speed should be traded off to boost the accuracy to increase the complexity for the DarkNet architecture. The most important feature of YOLOv3 is to make decisions at three different scales, and it is a completely convolutional network and generates output by 1×1 detection kernel.

References

1. YOLOv1: https://arxiv.org/abs/1506.02640.
2. YOLOv2: https://arxiv.org/abs/1506.02640.
3. YOLOv3: https://pjreddie.com/media/files/papers/YOLOv3.pdf.

4. DarkNet: https://pjreddie.com/darknet/.
5. He, K., Zhang, X., Ren, S., & Sun, J. (2016). Deep residual learning for image recognition. In *Proceedings of the IEEE conference on computer vision and pattern recognition* (pp. 770–778.
6. https://arxiv.org/pdf/1612.08242.pdf.
7. https://medium.com/@jonathan_hui/real-time-object-detection-with-yolo-yolov2-28b1b93e2088.
8. https://pjreddie.com/media/files/papers/YOLOv3.pdf.
9. Gordon, D., Kembhavi, A., Rastegari, M., Redmon, J., Fox, D., & Farhadi, A. (2017). Iqa: Visual question answering in interactive environments. arXiv preprint arXiv:1712.03316.
10. Huang, J., Rathod, V., Sun, C., Zhu, M., Korattikara, A., Fathi, A., Fischer, I., Wojna, Z., Song, Y., Guadarrama, S., & Murphy, K. (2017). Speed/accuracy trade-offs for modern convolutional object detectors. In *Proceedings of the IEEE conference on computer vision and pattern recognition* (pp. 7310–7311).
11. Krasin, I., Duerig, T., Alldrin, N., Ferrari, V., Abu-El-Haija, S., Haija, A., Kuznetsova, H., Rom, J., Uijlings, S., Popov, A., Veit, S., Belongie, V., Gomes, A., Gupta, C., Sun, G., Chechik, D., Cai, Z., Feng, D. N., & Murphy, K. (2017). Openimages: A public dataset for large-scale multi-label and multi-class image classification. *Dataset* available from https://github.com/openimages.
12. Lin, T.-Y., Dollar, P., Girshick, R., He, K., Hariharan, B., & Belongie, S. (2017). Feature pyramid networks for object detection. In *Proceedings of the IEEE Conference on Computer Vision and Pattern Recognition* (pp. 2117–2125).
13. Parham, J., Crall, J., Stewart, C., Berger-Wolf, T., & Rubenstein, D. (2017). Animal population censusing at scale with citizen science and photographic identification. In *2017 AAAI Spring Symposium Series.*
14. Redmon, J. & Farhadi, A. (2017). Yolo9000: Better, faster, stronger. In *Computer Vision and Pattern Recognition (CVPR), 2017 IEEE Conference on* (pp. 6517–6525). IEEE.
15. Ren S., He K., Girshick R., & Sun J. (2015). Faster r-cnn: Towards real-time object detection with region proposal networks. arXiv preprint arXiv:1506.01497.
16. Russakovsky, O., Li, L.-J., & Fei-Fei, L. (2015). Best of both worlds: human-machine collaboration for object annotation. In *Proceedings of the IEEE Conference on Computer Vision and Pattern Recognition* (pp. 2121–2131.
17. Shrivastava, A., Sukthankar, R., Malik, J., & Gupta, A. (2016). Beyond skip connections: Top-down modulation for object detection. arXiv preprint arXiv:1612.06851.
18. Szegedy, C., Ioffe, S., Vanhoucke, V., & Alemi, A. A. (2017). Inception-v4, inception-resnet and the impact of residual connections on learning. In *Thirty-First AAAI Conference on Artificial Intelligence.*
19. Cai, H., Wu, Q., Corradi, T., & Hall, P. (2015). The crossdepiction problem: Computer vision algorithms for recognising objects in artwork and in photographs. arXiv preprint arXiv:1505.00110.
20. Dalal, N., & Triggs, B. (2005). Histograms of oriented gradients for human detection. In *Computer Vision and Pattern Recognition, 2005. CVPR 2005. IEEE Computer Society Conference on* (Vol. 1, pp. 886–893). IEEE
21. Dean, T., Ruzon, M., Segal, M., Shlens, J., Vijayanarasimhan, S., Yagnik, J., et al. (2013). Fast, accurate detection of 100,000 object classes on a single machine. In *Computer Vision and Pattern Recognition (CVPR), 2013 IEEE Conference on* (pp. 1814–1821). IEEE.
22. Donahue, J., Jia, Y., Vinyals, O., Hoffman, J., Zhang, N., Tzeng, E., & Darrell, T. (2013). Decaf: A deep convolutional activation feature for generic visual recognition. arXiv preprint arXiv:1310.1531.

Chapter 10
IoT-Based Monitoring and Management of Electric Vehicle Charging Systems for DC Fast Charging Facility

R. Suresh Kumar, P. K. Rajesh, J. Joys Nancy, S. Abirami, and K. Vishnu Murthy

Introduction

The conventional powered internal combustion engine (ICE) emits toxic gases from the tailpipe, which leads to emissions and health risks associated with the respiratory system. In Europe, Norway aims to transition 25% of its vehicle to electric propulsion and 100% of passenger cars by the year 2030. Asian subcontinent countries such as India decided 100% electrification on two and three wheelers by the year 2025. Besides, vehicle populations are growing the demand for the charging infrastructure and energy storage requirements of facilities, which present the challenges faced by electrification.

However, with the current charging stations powered by alternating current (AC), a slow charging station typically needs an 8–12 h duration to charge 0–100% SOC of the EV's storage devices. During charging, EVs primarily require proper authentication and authorisation between the vehicle and charging station, security for payment, load balancing during peak and low utilized conditions [1].

The prime setback for the xEVs charging duration is with existing AC chargers, which hinders the implementation and adoption. To overcome these challenges, DC charging stations have been introduced by EV manufactures. The challenges associated with DC chargers align with battery pack health and vulnerability of safety due to quick charging. To overcome the above blockades, IoT integrated facilities need to transform efficiency and safety leading to a new paradigm shift. Presently available chargers are modified with a monitoring process for charging xEVs, where

R. Suresh Kumar (✉) · P. K. Rajesh
PSG College of Technology, Coimbatore, India
e-mail: rsk.auto@psgtech.ac.in

J. Joys Nancy · S. Abirami · K. Vishnu Murthy
Sri Krishna College of Technology, Coimbatore, India

© Springer Nature Switzerland AG 2020

G. R. Kanagachidambaresan et al. (eds.), *Internet of Things for Industry 4.0*, EAI/Springer Innovations in Communication and Computing,
https://doi.org/10.1007/978-3-030-32530-5_10

xEV vehicle parameters are articulated with a smartphone device assisted cloud computing facility. The emerging IoT driven EV charging stations are manufacturer assisted secured for the users against cybersecurity threats and issues [2–6].

The IoT bandwidth cloud promotes the offer to the applications to receive, analyse, control and backup the stored parameters to assist EV users to enhance safe and reliable operations. With the progression of IoT, insecurities and vulnerabilities of the systems were rectified with communication protocols adopted by integrated IoT fast chargers. Their implementation leads towards the adoption of IoT driven DC fast chargers with secure interconnection nodes (Fig. 10.1).

A sensor multiplex with network topology incorporation of IoT enabled devices in EV chargers abstracts the real-time data about the system supervision and performance. While supervising integration and deployment of IoT enabled DC charging infrastructure, it is commanding to recognise the challenges related to security issues. The IoT protocols must be subjective to develop secure, compact and decision-making DC chargers for EVs for quick charging functionality [6–11]. This chapter makes the effort to address an approach and a secure design for IoT integrated DC charging infrastructure. The chapter focuses on architecture to encompass safe-keeping and confidentiality of bill payments. Also, the chapter discusses how the IoT enabled DC fast chargers differ from the conventional AC charging infrastructure facilities. The EVs charged at different locations with standard charging devices and emphasis on implementing the supervisory charging environment between EVs and charging station with authentication integral is addressed.

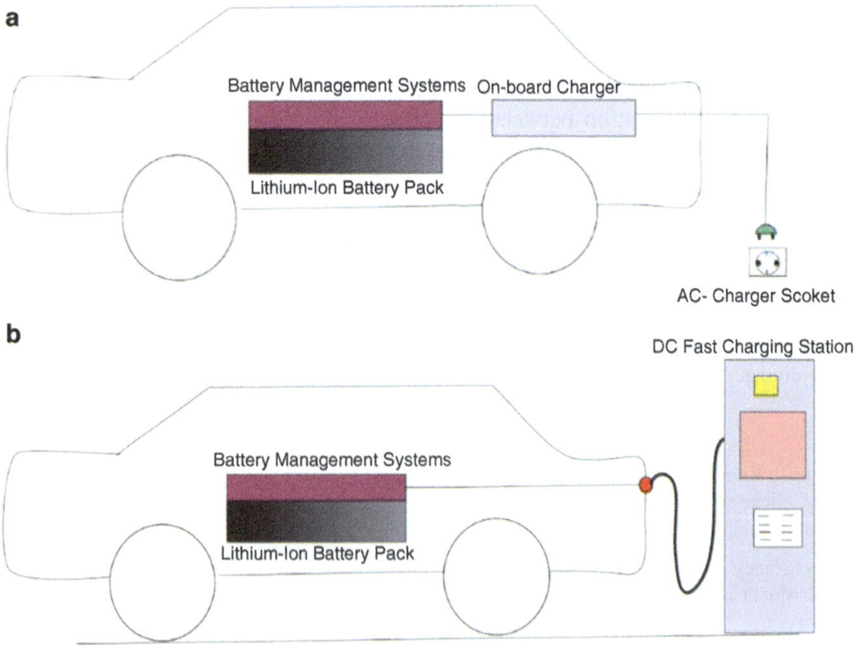

Fig.10.1 (**a**) AC charging systems. (**b**) DC fast charging stations

The abduction of information between chargers and electricity suppliers to the power grid are not emphasised in this chapter due to policy variation of the power transmission and distribution. The chapter discussions are made on analysis of the current scenario of EV charging infrastructure, methods to design a secure EV charging system, exploitation of relevant theory and risk assessment study. Finally, conclusions and certain limitations of the smart IoT driven charging infrastructure facilities are indicated [12–17].

Installed Present Charging Infrastructure Review

One of the topmost priorities in the smart city facility is to make efficient energy utilisation sustainable. Plug-in hybrid electric vehicles (xEVs) and plug-in Electric Vehicles (EVs) are the two main architectures manufactured by the EVs manufacturers. For the range and onboard utilisation, two powertrain deployment was the prime focus. EV using countries have EVs charged with AC and DC chargers based on controlled and adapter circuits with battery pack capacity. DC chargers stations expressly reduce the time by 80% depending on battery capacity [18]. DC chargers can charge the battery pack from zero to 100% state of charge (SOC) for an electric vehicle from 650–1850 seconds [16, 19–26].

Also, installation cost, power electronics components and time for developing IoT integration are the main challenges. Certain studies portray charging installations in the current charging stations lacked sacredness of personal information about the users. The vulnerabilities ranging from the ID for the vehicle manufacturers assigned by service providers might lead to encryption if the standard protocols such as OCPP are used. Consequently, there is also the opportunity to come up with new guideline protocols for the IoT fitted DC chargers.

The security process relevant to hacking and injection of malware being slow in updating is also a vital concern for the IoT enabled DC chargers. Unsanctioned admittance of fake EV user ID and illicit transactions from actual account holders are also critical challenges for the xEVs manufactures deployment. These key challenges are the prime focus that should be addressed by the manufactures at the initial phase of DC chargers installation.

The exploitation of wireless communication with IoT leads to a creative engineering industry on horizontal and different verticals. On the other hand, high-speed grid and soft computing devices become cheaper, which enforces EV charging infrastructure to proceed with intelligent chargers. Certain studies investigated IoT testing standards, such as physical security, data transfer in cloud, certification and firmware [26–31].

To overcome the above issues during the initial phase of product development, the EV and DC fast charger manufacturers need to address them. Security and discretion considerations scrutinised and validated by standard protocol promote the reliability of the charging infrastructure. This chapter provides insight into the IoT compelled rapid charging of battery packs with the secured baseline to overcome the threat of high steadfast DC chargers.

IoT-Based DC Fast Charging Deployment Requirements

The following factors are the requirements to deploy IoT-based DC fast chargers [32].

- The abstraction of xEVs user necessities with technological demands to estimate the latent attainment of IoT-based charging integrated infrastructure facility.
- Technology-driven factors and tasks that affect the implementation of DC fast chargers and with integrated IoT-based monitoring and control flow process.
- Validation of design and product development sequences by xEVs makers, EV manufacturers, internet service providers, component suppliers and investors.
- The collaboration of cloud computing service providers with power grid suppliers.
- Assessment of legislative guidelines, framework and investment activities influencing towards quick charging DC chargers deployment.
- Scrutinising vulnerability of energy storage devices in the DC fast charging facility remotely through cloud computing facility.

Also, a systems-level fault diagnostics study will be carried out on the EVs and the information will be shared with the service provider (service station) to detect and change the defect devices at the systems level to enhance the reliability of EVs. The communications to the grid should have sufficient intelligence to meet the different bandwidths of vehicle charging populations. The chargers are fused with data of the vehicle, charger status and substation. It enhances the effective operating mode of electric vehicles. Besides, if the demand is very minimal while the vehicle is on the charger, the charges tariff can discount to the consumer, which promotes electrification of transportation and simultaneously optimises the grid and generates station power.

IoT context with numerous features and advancements are applied to EV charging infrastructure facilities as a core design of information and communication technology. Also, concurrent growth in artificial intelligence, neural network, big data analytics, machine learning and cloud computing will permit controlling, processing and making judgments based on the enormous amount of information needed. A large number of devices and systems tangled will make improvements in machine-to-machine communications and will empower the required real-time bidirectional communications to meet the needs of security, reliability and quality of service (Fig. 10.2).

Owing to the constraint of the limited range and the long charging times, several techniques required to optimise the battery usage and to enhance the range of EVs have been developed. The two important investigation encounters that need to be addressed are: (1) Energy-efficient EVs routing charging to established search algorithms for the adoption of the demand for EVs to calculate the route's ability to enhance the driving range; (2) Battery efficiency, battery parameters and faculty condition, need to be monitored and data shared with the service and original equipment manufacturer for preventive maintenance to maximise battery health and life (Fig. 10.3).

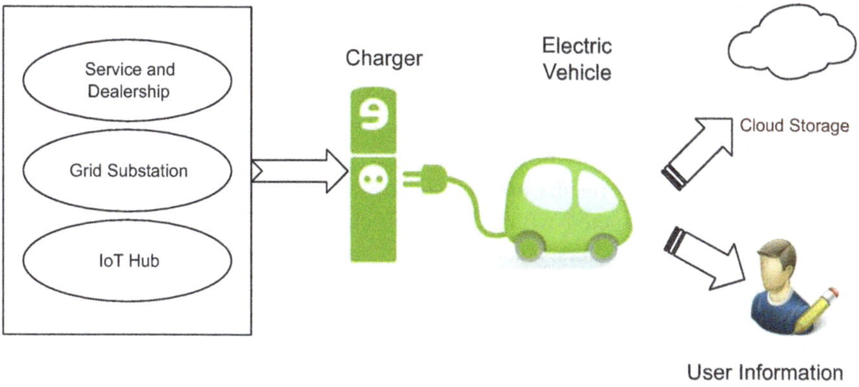

Fig. 10.2 Attributes of IoT-based DC fast charger systems

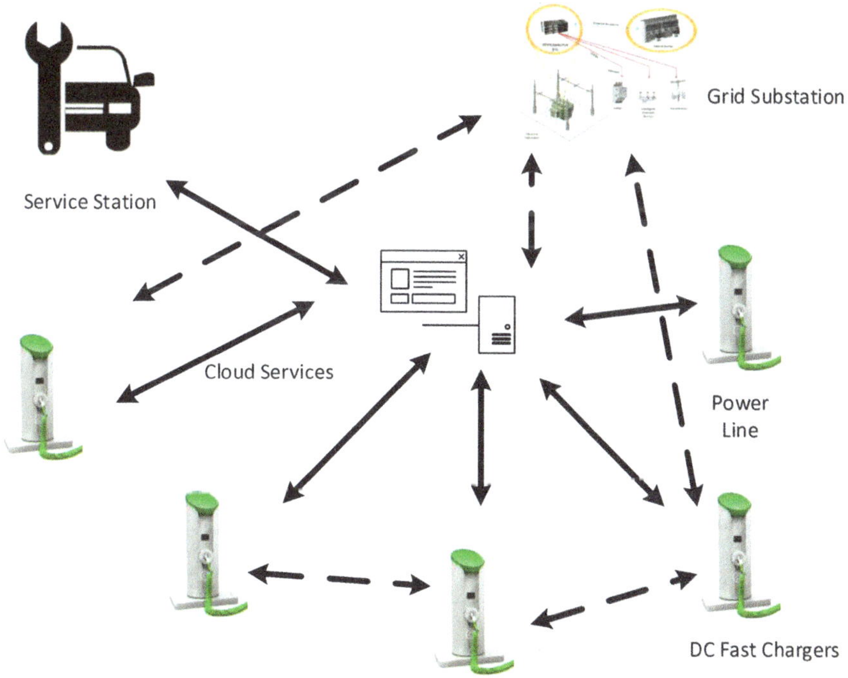

Fig. 10.3 Integrated service providers for DC fast charging systems

Electric Vehicle Charging

Battery Charging

The main challenges of EVs are charging and cost of the battery. Battery overcharging and deep discharge results in rapid ageing characteristics. The present energy storage devices constructed with lithium-ion chemistries are more vulnerable to various physical parameters, such as overcharged, temperature window, safety issues, state of charges, cycle life, shelf life and cell balancing. To overcome these challenges, sensors and estimators algorithms with actuators integrated with battery management and supervisory control termed as battery management systems (BMS) are needed. These BMS have many serviceable segments and key challenges, such as measurement of cell voltage to correlate state of charge, cell balancing, safe operating temperature and fault diagnosis. Thousands of cells are connected in series and parallel combinations as a module, and a few of the module integrations are said to be a vehicle battery pack. The battery pack and a motor controller are the EVs powertrain management systems. Proven literature demonstrates an internal resistance equivalent circuit (Rint) model where the charging and discharging have a similar trajectory. The equivalent circuit model (ECM) and the study of the Kalman filter helps to estimate the state of charge shown in Fig. 10.4. By using the Coulomb counting Ampere-hour integral method adopted to estimate SOC, the voltage of the cell to battery pack measurements correlated with SOC.

The xEVs battery pack state of charge (SOC) and state of health (SOH) predication provides the input BMS to display the remaining charge as monitoring the slowly varying battery ageing parameters. Most of the survey depicted that an adaptive extended Kalman filter (AEKF) is the best suited for the highly non-linear systems.

Certain research depicts the capacity estimation technique determined by the Q-estimation algorithm. The structure shown in Fig. 10.4 is a double layer estimation structure integrated with an algorithm for combined SOC and Q evaluation, and

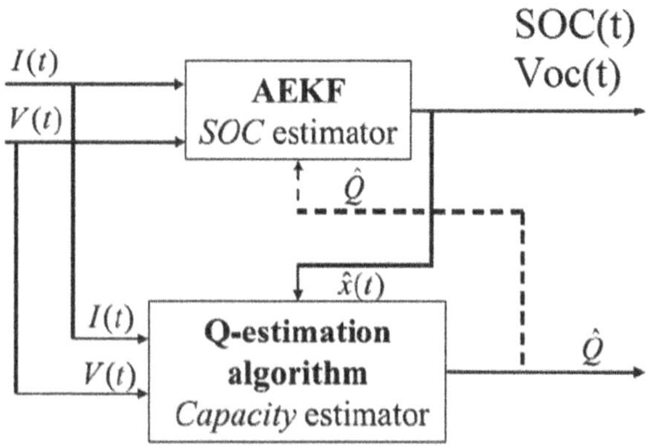

Fig. 10.4 Adaptive extended Kalman filter [33]

the SOC estimation is determined from an adaptive extended Kalman filter (AEKF). The Q-estimation algorithm provides the AEKF filter with an update value of the battery capacity, given the actual level of ageing. During fast charging, the Q value factor tunes the charger to set automatically based on the ageing.

Figure 10.5 depicts the charging and discharging profile of a lithium-ion cell constructed with ion phosphate (LFP) chemistry. The plot shows the hysteresis of the charging and discharging profile. The Q algorithm predicts the level of ageing under different operating temperature conditions. The collected data about the history of operation will provide the information to the user about the remaining useful life of the battery pack of the vehicle (Fig. 10.6).

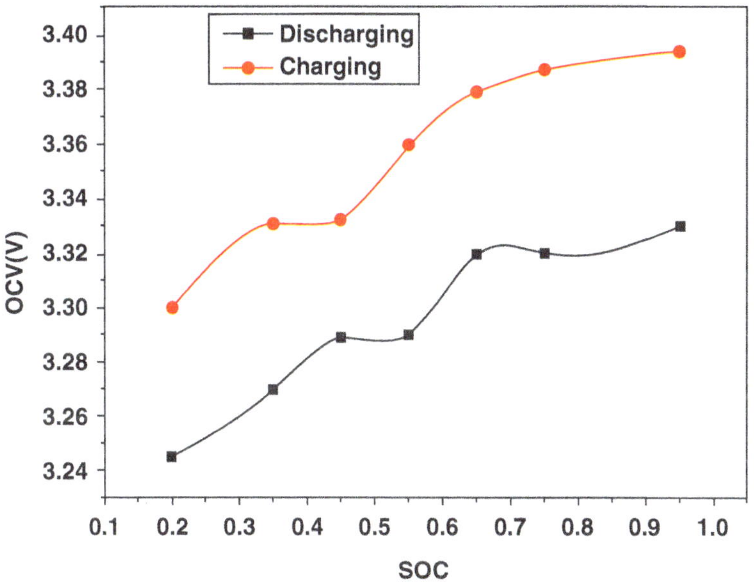

Fig. 10.5 Charging and discharging profile vs state of charge for 18,650 lithium-ion phosphate cell

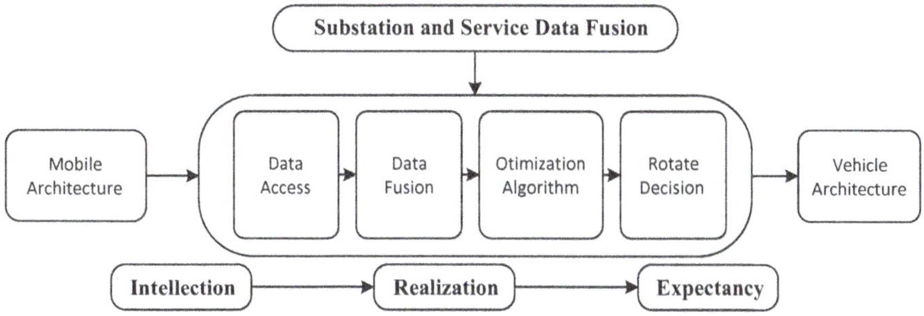

Fig. 10.6 Battery parameters observing system using IoT

The design facilitates the assimilation of the different sensors, regardless of their communication technology. It is because the block outfits encapsulation mechanisms for the data access received from different communication protocols through the wireless topology. In this manner, the design features can obtain information from different sensors connected to the system. The data obtained from the vehicle decides the decision for the charging station based on the availability from nearby charging stations. The next layer is to obtain information from any of the sensory networks. Data collected by the different sensors in the mobile architecture is pre-processed for data access. As the data from assorted sources receives information fusion, optimisation algorithms are performed to represent the knowledge for prediction and estimation for the demands. The optimisation algorithms' mechanisms depend on the implementation of the information superimposed in the data access. The optimisation model combines information from derived routes, on the geographic locations of the routes, earlier ingestion data, information from EVs sensors, instrument cluster devices and mobile navigation sensors. The optimisation models help sort levels of the traffic conditions, the route travelled, the battery level and the physical strain of the user. Certain optimisation works on the particle swarm optimisation (PSO) were synchronised by a neural network to optimise several constraints in the routes. It is expected that the control strategy with real-time optimisation estimates and predicts based on drive cycle demand and GPS data. From the optimal torque demand by the powertrain, the battery pack parameters adopt the actual torque by shifting torque demand to the road situation by limiting the speed. However, it relies on a high reckoning effort to encounter the real-time requirement. The fused sensor network deployed in the design is the most commonly used for electric vehicles and chargers the battery based on the optimised route location. The battery SOC, temperature, distance travelled and network faults in the battery pack will be crucial parameters to optimise battery stacking results as a cost structure of the intelligent systems.

Electric Vehicle Battery Management Using IoT

The IoT grounded battery-conditioning system for EVs discussed in this section comprises a communication protocol to the mobile architecture, data acquisition, cloud storage and an intelligent vehicle instrument cluster. An integrated IoT embedded architecture provides communication from the different data flow nodes. Encapsulated data is connected via the Internet gateway for all battery system parameters such that it enables one to back-up, analyse and control the critical parameters through the cloud computing systems.

The integrated IoT-based battery conditioning system communicates digitally with TCP/IP via Mat file (MATLAB) format for data acquisition purposes. The data from battery measurement parameters from the mobile architecture are extracted for further needs. Integrated protocol are optimised for connecting devices to vehicle operators, a special case of the D2S pattern, since vehicles are connected to the serv-

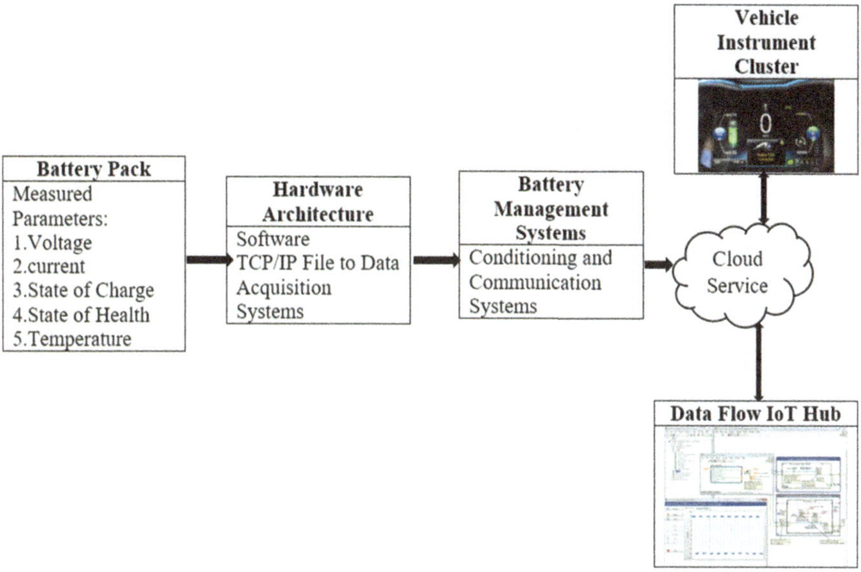

Fig. 10.7 Battery parameters observing system using IoT

ers through the internet data in the vehicle. Data is shared to the database and web server and client on the cloud system through the Internet gateway. The data in the cloud system is processed, analysed and a decision arrives at operators through the xEVs touch-based instrument cluster panel. The IoT architecture for the battery pack observing systems will be an embedded system utilised as an IoT to provide suggestions via hardware integration, for the main function to acquire data acquisition and as an Internet gateway for the data access. This acquired data is processed into an optimised algorithm estimation and a prediction is proposed using Kalman filter with the layout shown in Fig. 10.7, and the same is done using a graphical user interface to the data server for programming and executes in the vehicle instrument cluster.

The web-based GUI executes the data acquisition functions to regain battery measurement parameters from the MATFILE format. Transferred data through structured query language will be stored in cloud computing stations.

IoT Charging Station Integration Requirements

In the designing process of IoT integrated xEV charger systems the following requirements must prevail for safety and security. Guidelines and technical measures correlation expose threats in the present chargers. These procedures and technical scrutiny aim to alleviate extortions, liabilities and threats identified in the IoT driven EV chargers (Table 10.1).

Table 10.1 Standard for DC charging station and EV connectors and the communication standard between EV and charging standard [1]

Communication	Standard	Description
DC electric vehicle charging station [34]	IEC 61581-23	Electric vehicle conduction-DC EV charging station
Plugs, socket outlet, vehicle connectors and vehicle inlet for DC charging [35]	IEC 62916-3	Dimensional compatibility and interchangeability requirement for DC and AC pin and tube type vehicle connectors
Communication between plug-in vehicles and off-board DC chargers	SAEJ2847/2	Requirement and specifications for communication between plug-in vehicles and off-board DC chargers
Communication between EV	ISO 15118	Vehicle to the grid communication interface
DC charging specification communication	DIN 70121:2014:2	Digital communication between xEVs and deployed DC chargers [1].

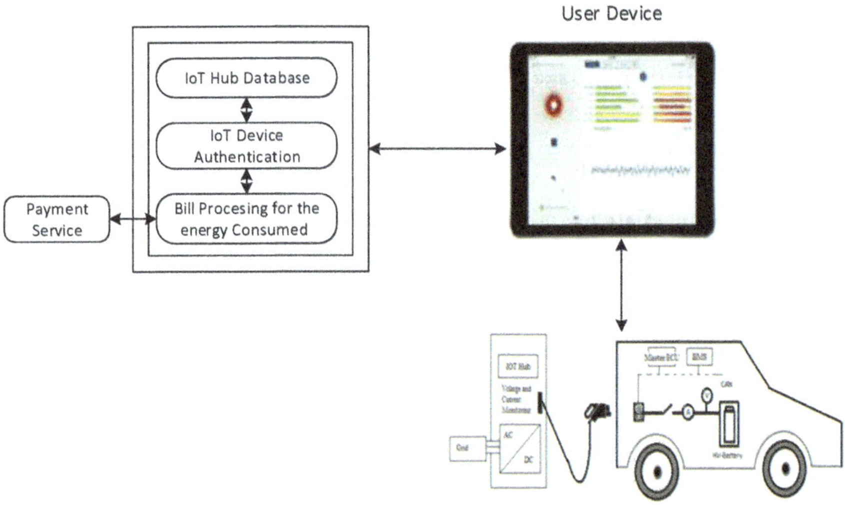

Fig. 10.8 A typical IoT charging station integration

Figure 10.8 depicts IoT modem connections to the cloud service, and it shows the communication and power line for the rapid DC chargers. Mobile applications installed in the xEVs' user device ensure the security and authentication for the start of charging. In this schematic, the user's locations are identified via navigation systems with charger ID for the service provider to assess the vehicle required demand. Also, the user's requisition for charging station ID must be entered. If the user ID is matched and further processed, a payment service on the cloud service initiates a payment transaction with an external online payment service provider from wallet, a credit card, a debit card, RFID-based detection systems or net banking payment. After the payment transaction is authorised by the payment service, the cloud service transmits a command to the IoT hub in fast DC chargers to authorise the connected EV of the user and allow the use of the charging station. The payment is only for the consumed energy initially calculated based on the SOC of the battery. As per the standard DIN SPEC 70121 and ISO 15118 bidirectional communications are initiated between the vehicle and cloud computing to the user via the user display devices (smartphone app). Also, the payment service is a bank account that is interlinked with cloud service for the cost deduction authentication. If the users require a partial charge to charge the battery and want to travel a shorter distance, the transaction money refund to the user account is also one of the potential benefits from the IoT enabled DC fast chargers.

Conclusion

In rapid development, IoT technology drives opportunities for xEV manufactures to privilege their strong presence in the market. Besides, IoT still lacks platform standards, network standard protocols, privacy, interoperability and faces malware threats. IoT manufacturers have provided solutions and consideration for the design and development of highly secure authentication.

Also, the integrated IoT should possess security data on areas that are open to exploitation and have it critically assessed before the integration. During the design process itself, essential checkpoints in architecture are thoroughly analysed with threats associated in the setup. Additionally, the charging infrastructure of xEVs integrated with different xEVs manufactures is going to adopt the standard protocol. The main factors in portraying successful deployment are the recognition, detection, access management, encryption, adaptability and safe transactions for bill payment, applied to every entity of the IoT driven charging infrastructure. A test of the smart application in the user interface display or mobile enables the users to monitor and control parameters of the EVs for the knowledge and information about the charger systems utilisation. Cloud information of the individual EV database will be helpful to the EV manufacturers for research and development details. Also, the same details will provide a clear usage foundation of the batteries and charge/discharge patterns.

During peak and low demand conditions, the energy providers can analyse the consumption of energy. Also, the consumed energy and charging time through IoT provides data about the best EVs for the customers to choose the right vehicle. Through the IoT self-discharge and ageing characteristics of the energy storage devices, information can be predicted for certain vehicles not operated for a couple of days or longer duration.

References

1. Ranabhat, P. (2018). Secure design and development of IoT enabled charging infrastructure for electric vehicles; using CCS standard for DC fast charging. Metropolia University of Applied Sciences, Information Technology.

2. International Electrotechnical Commission. (2014). *IEC 61851-23. Electric vehicle conduction system – Part 23 – D.C. electric vehicle charging station*. Geneva: International Electrotechnical Commission.
3. Shaffers, H., Komninos, N., Pallot, M., Trousse, B., Nilsson, M., & Oliveira, A. (2011). Smart cities and the future internet: Towards cooperation frameworks for open innovation. In *The Future Internet* (pp. 431–446). Berlin/Heidelberg: Springer.
4. Zhang, Q., Liu, M., Lin, X., Liu, Q., Wu, J., & Xia, P. (2019). Optimal resonant beam charging for electronic vehicles in the internet of intelligent vehicles. *IEEE Internet of Things Journal, 6*(1), 6–14.
5. Wang, H., He, L., Qi, X., Wang, H., Li, G., & Dai, X. (2012). Simulation system of the performance of power battery for electric vehicle based on internet of things. In *Proceedings of 2012 International Conference on Measurement, Information and Control* (Vol. 2, pp. 681–684).
6. Ouya, A., De Aragon, B. M., Bouette, C., Habault, G., Montavont, N., & Papadopoulos, G. Z. (2018). An efficient electric vehicle is charging architecture based on LoRa communication. In *2017 IEEE International Conference on Smart Grid Communications (SmartGridComm)* 2017, vol. 2018–January, no. October, pp. 381–386.
7. Li, H., Zhang, W. S., Shi, J. Y., & Ma, X. Y. (2013). The sensing technologies in the electric vehicle charging and exchanging station. *Applied Mechanics and Materials, 303–306*, 1456–1459.
8. Kobeissi, A. H., Bellotti, F., Berta, R., & De Gloria, A. (2018). IoT Grid alignment assistant system for dynamic wireless charging of electric vehicles. In *2018 Fifth International Conference on Internet of Things: Systems, Management and Security*. IoTSMS, 2018 (pp. 274–279).
9. Karpenko, A., Kinnunen, T., Framling, K., & Dave, B. (2018). Open IoT ecosystem for smart EV charging. In *2018 Global Internet Things Summit (GIoTS)* 2018.
10. Li, K., Xue, Y., Cui, S., & Niu, Q. (2014). Intelligent computing in smart grid and electrical vehicles: International conference on life system modeling and simulation, LSMS 2014 and international conference on intelligent computing for sustainable energy and environment, ICSEE 2014 Shanghai, China, September 20–23, 2014 Proceedings, Part III. In *Commun. Comput. Inf. Sci.*, Vol. 463, no. September 2014.
11. Hussain M., Alam M. S., & Beg M. M. S. (2019). *Proceedings of 2nd International Conference on Communication, Computing and Networking*, Vol. 46. Springer Singapore.
12. Hariri A. O., El Hariri M., Youssef T., & Mohammed O. (2018). A decentralized multi-agent system for management of en route electric vehicles. In *Conf. Proc. – IEEE SOUTHEASTCON* (Vol. 2018, pp. 1–6).
13. Bahadoorsingh, S., Meetoo, C., Sharma, C., & Hosein, P. (2019). A unique approach to demand side management of electric vehicle charging for developing countries. In *2018 IEEE International Smart Cities Conference (ISC2)* (pp. 1–2). IEEE.
14. Asaad, M., Ahmad, F., Alam, M. S., & Rafat, Y. (2018). IoT enabled monitoring of an optimized electric vehicle's battery system. *Mobile Networks and Applications, 23*(4), 994–1005.
15. Ruben, A., Vitor, M., Joao, C., Andres, A., Joao, L., & Jose, A. (2018). Development of an IoT system with smart charging current control for electric vehicles. *IECON 2018 - 44th Annual Conference of the IEEE Industrial Electronics Society* (pp. 4662–4667). IEEE.
16. Chaudhary, R., Jindal, A., Aujla, G. S., Aggarwal, S., Kumar, N., & Choo, K.-K. R. (2019). BEST: Blockchain-based secure energy trading in SDN-enabled intelligent transportation system. *Computers & Security, 85*, 288–299.
17. Hua, M., Li, C., Huang, Y., & Yang, L. (2017). Throughput maximisation for UAV-enabled wireless power transfer in the relaying system. In *2017 9th International Conference on Wireless Communications and Signal Processing (WCSP)* (pp. 1–5). IEEE
18. Guide to EV charging Jun-2018. https://www.zap-map.com/charge-points/
19. Meng, Q., & Jin J. (2011). The terminal design of the energy self-sufficiency Internet of Things. In *2011 International Conference on Control, Automation and Systems Engineering (CASE)* (pp. 1–5). IEEE.

20. Zhang, R., Cheng, X., & Yang, L. (2016). Flexible energy management protocol for cooperative EV-to-EV is charging. *IEEE Transactions on Intelligent Transportation Systems, 31*(3), 97–101.
21. Zeng, F., Zhang, R., Cheng, X., & Yang, L. (2017). Channel prediction based scheduling for data dissemination in VANETs. *IEEE Communications Letters, 21*(6), 1409–1412.
22. Barbarisi, O., Vasca, F., & Glielmo, L. (2006). State of charge Kalman filter estimator for automotive batteries. *Advances in Automotive Control-Control Engineering Practice, 14*(3), 267–275.
23. Huvio, E., Grönvall, J., & Främling, K. (2002). Tracking and tracing parcels using a distributed computing approach. In O. Solem (Ed.), *Proceedings of the 14th Annual conference for Nordic researchers in logistics (NOFOMA'2002), Trondheim, Norway* (pp. 29–43).
24. Bellavista, P., Cardone, G., Corradi, A., & Foscini, L. (2013). Convergence of MANET and WSN in IoT urban scenarios. *IEEE Sensors Journal, 13*(10), 3558–3567.
25. Bellavista, P., Cardone, G., Corradi, A., & Foscini, L. (2013). Convergence of MANET and WSN in IoT urban scenarios. *IEEE Sensors Journal, 13*(10), 3558–3567.
26. Shaffers, H., Komninos, N., Pallot, M., Trousse, B., Nilsson, M., & Oliveira, A. (2011). Smart cities and the future internet: Towards cooperation frameworks for open innovation. In *The future internet assembly* (pp. 431–446). Berlin, Heidelberg: Springer.
27. Hernandez-Munoz, J., Vercher, J., Munoz, L., & Galache, J. (2011). "Smart cities at the forefront of the future internet", the future internet, lecture notes computer. *Science, 6656*, 447–462.
28. Mulligan, C. E. A., & Olsson, M. (2013). Architectural implications of smart city business models: An evolutionary perspective. *IEEE Communication Magazine, 51*(6), 80–85.
29. Meetoo, C., Bahadoorsingh, S., Balbadar, V., Jaglal, D., Sharma, C., Baboolal, K., & Williams, M. (2018). Electric vehicle policy formulation framework for SIDS in the Caribbean. In *2018 IEEE Transportation Electrification Conference and Expo (ITEC)* (pp. 152–157). IEEE.
30. Mahadeo, G. E., Bahadoorsingh, S., & Sharma, C. (2017). Analysis of the impact of battery electric vehicles on the low voltage network of a Caribbean island. In *2017 IEEE Transportation Electrification Conference and Expo (ITEC)* (pp. 364–369). IEEE.
31. Buratti, C., Conti, A., Dardari, D., & Verdone, R. (2009). An overview on wireless sensor networks technology and evolution. *Sensors, 9*(9), 6869–6896.
32. Kobeissi, A. H., Bellotti, F., Berta, R., & De Gloria A. (2018). IoT grid alignment assistant system for dynamic wireless charging of electric vehicles. In *2018 Fifth International Conference on Internet of Things: Systems, Management and Security* (pp. 274–279). IEEE.
33. Taborelli, C., Onori, S., Maes, S., Sveum, P., Al-Hallaj, S., & Al-Khayat, N. (2016). Advanced battery management system design for SOC/SOH estimation for e-bikes applications. *International Journal of Powertrains, 5*(4), 325–357.
34. Fluhr, J., & Lutz, T. (2011). Chapter 8 communication with and for electric vehicles. InTech.
35. Morris, B., Foiadelli, F., Longo, M., & Zaninelli, D. (2015). e-Mobility forecast for the transnational e-corridor planning. *IEEE Transactions on Intelligent Transportation Systems, 17*(4), 680–689.

Chapter 11
An Evolution of Innovations Protocols and Recent Technology in Industrial IoT

V. Saranya, M. J. Carmel Mary Belinda, and G. R. Kanagachidambaresan

Introduction

Manufacturing industries experience complex difficulties in building supply chain connection in this technological business world. To develop a successful system, new concepts such as Internet of Things (IoT), Digital Technology, and Cyber Systems have gained more importance in the industrial sector including manufacturing units. The evolution in the digital era defines the Fourth Industrial Revolution (Industry 4.0) and it is also recognized as German high-tech strategy for future manufacturing industries [1]. Industry 4.0 produces overwhelming results by transforming the manufacturing and production process of industries. In addition, Industry 4.0 will play a vital role in transforming well-established companies into Smart industries with the help of Internet of Things (IoT) and Cyber Systems. A distributed approach produces an immense value in Industry 4.0, which highlights self-determining organization of processes and smart objects throughout the network by collaborating on the processes of real and virtual world [2]. The development of incorporated processes and human-machine communication motivate involvedness and liveliness for data transmission between industrial chains [3]. With the assistance of Industry 4.0, industries gain lot of operational efficiency in time, cost, and production. Designing and developing the infrastructure of IoT offer common platforms via cloud systems between associates in supply chains; consequently, business processes get optimized [4]. The notion of Industry 4.0 has gained great significance in the recent years. The increase in the usage of computerized

V. Saranya (✉)
Department of IT, Sri Krishna College of Engineering and Technology, Coimbatore, India

M. J. Carmel Mary Belinda · G. R. Kanagachidambaresan
Department of CSE, Vel Tech Rangarajan Dr Sagunthala R&D Institute of Science and Technology, Chennai, India

© Springer Nature Switzerland AG 2020

G. R. Kanagachidambaresan et al. (eds.), *Internet of Things for Industry 4.0*,
EAI/Springer Innovations in Communication and Computing,
https://doi.org/10.1007/978-3-030-32530-5_11

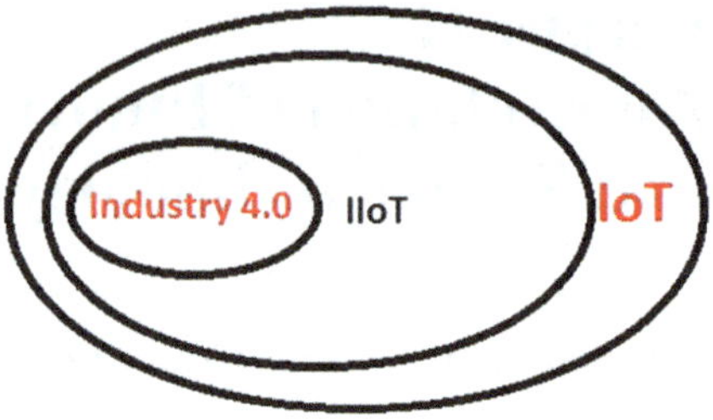

Fig. 11.1 Evolution of technology

systems after the Third Revolution, Industry 4.0 deals with developing more digitized systems and network incorporation via smart systems. Through Industry 4.0, smart systems would facilitate the substitution of human beings in assured tasks and effortlessness in the working atmosphere.

As IIoT features are similar to that of IoT in the industrial sector, and Industry 4.0 concepts are effectively subsets of IIoT, as shown in Fig.11.1, in identifying the difference between IoT and IIoT. Even though the fundamental concepts of both are the same, i.e., interrelated smart devices that permit remote sensing, data collection, processing, monitoring, and control are the parameters that recognize the IIoT as a subset of IoT, there are the strong necessities for constant operation and safety as well as the equipped technology employed in the industrial sector. Therefore, this chapter includes an evolutional study of IIoT in Section "Evolution of IIOT". Section "Innovations and Recent Technologies in IIoT" evaluates the innovations and recent technologies used, Section "Protocols in IIoT" discusses the protocols used for designing this environment, and finally Section "Conclusion" concludes this chapter.

Evolution of IIOT

The Industrial Internet of Things (IIoT) is a commercial technology, and because of its major benefits, it occupies a major part in the development of an industry. These technological developments have made innovations in the implementation of smart factories and predictive technology. As the machines are equipped with sensors, industries train the employees on the supply and delivery chains with the specific industrial tools to monitor and reply the output from the sensors and companies can be able to rationalize and restructure the industrial business operations. The Industrial Internet of Things (IIoT) discusses how the interconnected sensors, instruments, and other devices are interrelated together with workstations for industrial applications, including manufacturing and energy management. This connectivity permits data collection, exchange, and analysis, theoretically facilitating developments in efficiency and productivity as well as other economic benefits. The IIoT is an evolution of a distributed control system (DCS) that allows for a automation by using cloud computing to improve and enhance the method controls. The big data and analytics are preliminary to influence the end users.

IIoT is flagging the method for a broad change of predictable business processes from gathering to delivery to preservation to fabrication. Warehouse sensors permit firms to monitor stock cleverly so that shares and supplies are well-organized whenever needed. This prevents overstocking and under stocking of essential items. Sensors along the delivery chain make it likely to track shipments from the moment they leave the factory premises to the moment they arrive at the customer premises. Sensors along the production line lead to timely detection of potential breakdowns. By depending on projecting care to fix problems before they happen, companies sidestep costly interruptions and disruptions in production. All these exhibitions improve efficiency, minimize unnecessary expenses, and maximize quality.

- Contemporary Responses and Mass Facilities

IIoT has designed an alteration in the organization and updated expectancies. The smooth access up to the moment records, products, and services that our mobile devices provide us has caused an improved demand for real-time systems and responses, which is even more common in IIoT. As sensors provide present perception and statistics about warehouses and machines, a greater need for immediate, or real-time, services has arisen. The purpose is obvious: IIoT, this connection of business machines and sensors up the internet and different devices, is useless without mechanisms in area that react up to the moment notifications and updates. There is no experience in understanding about forthcoming issues or shortages if there is no manner up to date respond with the necessary speed and accuracy that avoid potential issues.

This shortage has marked a new beginning to an entire different innovation: crowd offerings. Businesses pool collectively all their assets—employees, partners, subcontractors, freelance professionals, and specialists—to create a pool of available service vendors to respond to the elevated demand generated through actual-time services. This crowd is based on discipline provider control software program and their mobile devices to live informed of service requests, product information, consumer records, and more.

Industry 4.0

The industrial IoT 4.0 is a suitable method for processing data such as sensor output, consumer input, identification of the service provider availability, knowledge provider, and other various effective real-time responses given by mobile devices and also gives an idea about the next step to industrial revolution. As the evolution of IoT to IIoT after twenty years created a potential evident in the new industrial technology. To imagine the continuous change in technology and redefine the daily lifestyle and work over the course made a remarking foundation in the field of IIoT. The connections between devices and accumulation of raw data with the help of Big Data Analytics made the industries to rely on artificial intelligence (AI) and edge computing for processing and analysis.

Revolution of IoT

In the revolution of IoT, Kevin Ashton created awareness that physical objects can be linked to a network during 1999. By this it is known that information to computers no longer needed to be programmed and given in the form of functions but sense the world around them by fingers without any added information given by the humans this process created a new computational technology era. IoT has evolved in all the perspectives from GPS to smart homes and smart cars to health and fitness monitoring and other smart environments and also sets a stage for the next industrial revolution by giving way to IIoT.

As an example, a smart home owner controls his home devices by working on an app on his mobile by just making some clicks and swipes for switching on the AC or heater and also checks the status of the refrigerator for pickup things from supermarket. Other apps from the mobile could be able to update the real-time status and current scenario around their living environments. The smart application system software is developed to perform the user-defined task much faster and controls the devices automatically by connecting with internet. The inputs and outputs from the devices can be viewed by using electronic gadgets, and also the overall process can be analyzed. Built-in sensors are used to make a trustworthy connectivity between devices for better determination of data and sequence of transmission while performing the required task.

Versions of IIoT

In the past centuries, goods and other necessary products are manufactured by using hand or by using animals. However, at the beginning of the nineteenth century manufacturing and production drastically introduced industry 1.0 and other operations rapidly developed based on the requirements [5]. The different versions of IIoT are listed as follows.

Industry 1.0
During the 1800s, water- and steam-powered machines were developed to aid the workers who started the industrial revolution 1.0 as the production capabilities increased and also business grew to a next level where the owners, managers, and other employees serving customers used steam-powered engines and water as a source of power. This led to a lot of changes in textile industries and constituted a major part of the British economy.

Industry 2.0
At the beginning of twentieth century, electricity became the primary source of energy for functioning the machineries. The users felt easier to use when compared with water- and steam-powered individual machines. Eventually, new machines were designed to use electric power sources and also with portability. To increase

the efficiency and effectiveness in manufacturing units, this period created a drastic development in major management programs where labors works for a part of job by increasing the productivity. The concept of assembly lines increased mass production. Lean manufacturing principles further enabled manufacturing industries to improve the quality and also produce good products.

The electrified machines in factories massively increased the production rates. The mass production from steel industries helped introduce railways into the system. Innovations in chemical industries lead to the invention of the synthetic dye, which marked a primitive state in the fabric design with various colors.

Industry 3.0

After a few decades of the twentieth century electronic devices using transistor, integrated circuits chips made automated individual machines to supplement the use of operating workers. Later, developments were made in the software side to control the electronic hardware. By this embedded mechanism, the components are integrated where material planning is made by using enterprise resource planning tools where humans can plan, schedule, and track products from industry to the customer. The concept of supply chain management formalized the result of extended geographic dispersion.

As the third revolution resulted in huge development in computers and information and communication systems, the industries lean on the digital technology in production that offered a Digital Revolution, which led to a change from analog and mechanical systems to digital systems. The digital era is also known as Information Age

Industry 4.0

In the twenty-first[1] century, the industrial revolution 4.0 gets connected with IoT using manufacturing techniques to enable systems for sharing information and analyze and use it for performing intelligent actions. It creates a cutting-edge technology in additive manufacturing, robotics, artificial intelligence and other cognitive technologies, advanced materials, and augmented reality, according to the article [6].

The development of new technology Industry 4.0 was first developed during the later stages of the twentieth century. Manufacturing execution systems, shop floor control, and product life cycle management were farsighted concepts that lacked the technology needed to make their complete implementation possible. Although "Industry 4.0" is the common term referring to the fourth industrial revolution, academics still argue to appropriately define the methodology (Fig. 11.2).

Applications of IIoT

1. Industrial Automation

Industrial automation is the most substantial and common application of IoT. Automation systems make machines and tools to function in an efficient way

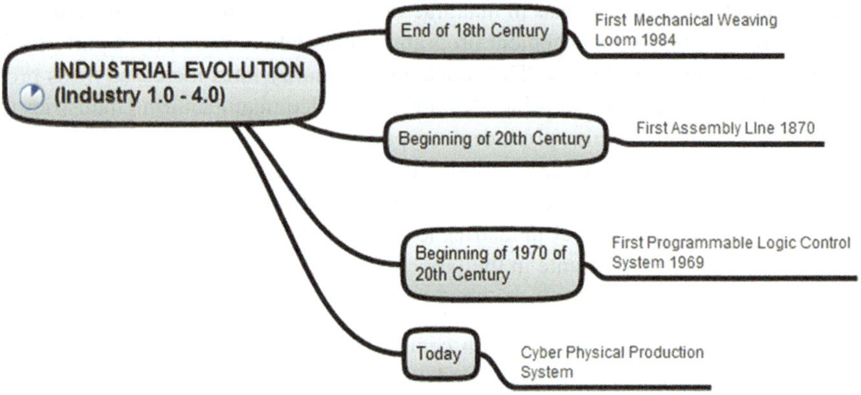

Fig. 11.2 Industrial evolution

by using sophisticated software tools that act as an observer and make developments for succeeding process rehearsals. A greater level of improvement has been shown by machine automation with accuracy. Automation tools such as PLC (Programmable Logic Control) and PAC (Programmable Automation Control) use smart sensor networks associated with a central cloud system, which collect huge amount of data. Application-oriented software are designed to analyze the data that improves accuracy, efficiency, error control, and easy access and reduces man power in specific tasks in industrial automation. The industrial IoT provides solutions in connecting industries with major components such as machines, tools, and sensors for network connection. These components provide effective solutions to check the status, shipment, schedule maintenance, and also monitor the process flow overtime.

2. Smart Robotics

Intelligent robots are developed by the industries based on the required specification and are created by IoT-enabled industries. By utilizing these robots, manufacturing units ensure smooth handling of tools and materials with efficiency and accuracy. Robots are designed to perform complex tasks with high embedded sensors that analyze real-time environment as an interface for man-machine interaction. Robotics networks are connected to cloud in a secure way for monitoring and controlling.

Engineers and designers can be able to access and analyze the data to yield quick actions for product enhancements and also prevent any unexpected failure due to machine failure.

3. Predictive Maintenance

Modern industrial machines are equipped with smart sensors for continuous monitoring of the current status of major components for detecting any critical issues before the system gets completely failed. The smart sensors notify even the maintenance warning to the centralized system and also send alert messages to the responsible teams for further processing. For every maintenance procedure, engineers are allotted with an analyzed schedule maintenance plans which do not affect the routine task and also unnecessary problems in the production line. Some unexpected machine failure may happen which leads to damage of products, delay in delivery, and also loss in business for producers. But in recent years, the development of technology has made maintenance of machines on real-time basis by storing the overall performance to a cloud system. History of each machines helps to process a predictive maintenance system.

4. Integration of Smart Tools/Wearables

Workforce can perform its task with improved accuracy and efficiency by integrating smart sensors in tools and machines. Specially planned wearables and smart glass support workforces to reduce error and expand safety at the functioning environments. Smart wearables can be able to intimate the users by giving instant alert messages to employees during critical situations such as gas leak or fire. Wearables can be designed for monitoring health conditions of individuals continuously.

5. Smart Logistics Management

Logistics is one of the major areas in many industries which needs continuous development to support increasing demands. Smart technology provides a perfect solution to solve complex operations and manage the delivery of goods to the customers efficiently. Marketing monsters such as Amazon uses drones to deliver goods to their customers. Advanced technologies such as drones offer better efficiency, accessibility, speed and require less manpower. However, initial investments are huge compared to conventional methods and implementation has limitations.

Airline is an added major industry that uses IoT for performing its daily operations at the manufacturing and predictive maintenance of planes in service. At the manufacturing plant, airline companies use IoT solutions to track thousands of components required for every single day at work. Unified management of portfolios helps to manage its supplies effortlessly.

6. Software Integration for Product Optimization

Smart analytics of any IoT system enhances the possibilities of the system for optimization and improvement. Many industries and major companies implement a customized software for deep analysis of huge data collected by the sensor networks and machines. The analyzed data gives a better overview of process improvement strategies for optimizing the products. The analysis of data and its behavior patterns provide cost-effective solutions in a period of time. As though analysis of large amount of data was very hard, inaccurate and more time consuming before the introduction of software tools.

7. Smart Package Management

The manufacturing units deliver its products without any damage by using IoT technology for package management, which gives a lot of convenience and efficiency. The smart sensors embedded on the packages could be able to detect the vibrations, atmospheric conditions such as temperature and humidity, and also they can be able to monitor the stages of packages and update the status periodically. Finally, the product delivery to customers is confirmed by sending messages.

8. Autonomous Vehicles

Industries use automotive vehicles using the IoT technology by enabling self-driving system to carry goods within the company premises. These vehicles can be able to detect obstacles along its path and find another route to reach its prescribed destination using the GPS technology and wireless technology to communicate with the control station [7].

Pros of Industrial Internet of Things

- Enhanced accuracy
- Optimization of product and process
- Analytical preservation and exploration
- Advanced competence
- Distant accessibility and observing
- Improved security
- Scalability of network
- Reduced energy consumption

Challenges in IIoT

Business becomes a constant battle for balancing the benefits of new technology with increased risks using sophisticated data. However, things are different in the case of critical infrastructure. Before adopting the IIoT with smarter components, it is wise to think about some challenges to overcome with efficient benefits.

1. *Security: Major Concern*

Increased usage of digital connectivity in IIoT deployment sets to lot of security risks. As the machines and other collected data are placed in a common cloud environment, it leads to higher risks creating cyber attacks against the critical infrastructure. Even though the records are maintained in a common repository, all the security measures have to be placed by using protocols.

2. *Preserving Visibility of "Things"*

Tracking of things make network engineers to build systems that show network visibility where the intruders can be able to easily analyze the valuable goods by getting a clue from either the customer or the suppliers. But the network could get started from the initial zero to potentially thousands of connected sensors and devices. It is really very hard to manage visibility for a single factory floor, but the Industrial Internet of Things will add additional challenges as you scale it globally.

3. *Protecting the Industrial Ethernet*

The industrial Ethernet creates a highway for building an IIoT environment where 100% internet connectivity is needed to perform each task. Hence, maintenance of systems can be made only through online services, but to go offline for maintenance can be guaranteed by using proper cables only for permanent usage with a logic of no data loss even during the connectivity issues.

4. *Unified Legacy Systems and IIoT Infrastructure*

A unified legacy system environment using IoT infrastructure focuses less on security and more attention on performing optimizing operations and third-party integrations. But in case of IIoT environment, a new modernizing infrastructure is developed by adding smart technologies, which focuses individually on security, third-party intrusion, and on all stages of processing.

5. *Data Storage*

As large amount of data being generated from different sources for any analysis to be made ensures rapid development of IIoT. All the raw data cannot be stored in the IIoT devices due to its less memory size and limited energy level. If the network is deployed in an unreachable or unreliable area, it may cause vulnerable threats leading to computational problems. To avoid the challenging aspects such as data processing, secure data storage, efficient data retrieval, and dynamic data collection in IIoT can be solved by designing a flexible and economic framework using for computing and cloud computing techniques.

Innovations and Recent Technologies in IIoT

The new innovations and implementation of recent technology have made the IIoT industries to enhance their process by increasing the production, marketing, and other business fields [8]. The new innovations can be able to provide some list services to the industrial group.

Collaborative Enterprise
In the current scenario, companies undergo high pressure from customers, stakeholders, and global competitors. To improve the performance, there must be an operational improvement focusing on business process and collaborative concepts. A collaborative management model (CMM) provides information on the complex interactions, applications, collaborations, and processes that an enterprise demands. This framework model is a basis for organizing and controlling the key business processes in an enterprise (Fig. 11.3).

Cybersecurity for Industries and Smart Cities
Cybersecurity secures the information and conditions transmission through the network. An effective integration of other technologies creates its own boundaries increasing the reliability. Cybersecurity concerns are obstructing determinations to

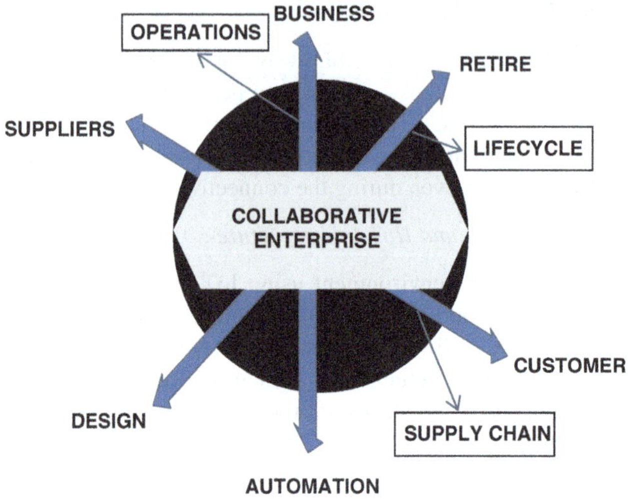

Fig. 11.3 Collaborative enterprise

implement new approaches for lower costs and more-competitive business approaches [9–13]. Cybersecurity Advisory Service can help to address the challenges through perceptive reports, individual consultations with experienced analysts, and peer networking opportunities. This provision provides an ultimate way to learn what others are doing, how they are doing it, and the benefits they are achieving. It also enables the consumers to keep updated on the several new developments occurring in cybersecurity technologies, practices, and principles (Fig. 11.4).

Digital Transformation

A successful digital transformation promotes an organization in a process that transforms its current state to a desired future state. In case of production operations, there involves moving from a local, soloed, procedural operating model to a more comprehensive model that is integrated to the enterprise, responsive to customers and market changes, and optimized in the context of overall business performance. The transformation is made based on the following (Fig. 11.5):

1. Customer-focused Innovation
2. Transformation framework to digital transformation
3. Technologies, security, and architecture
4. Business justification and solution selection

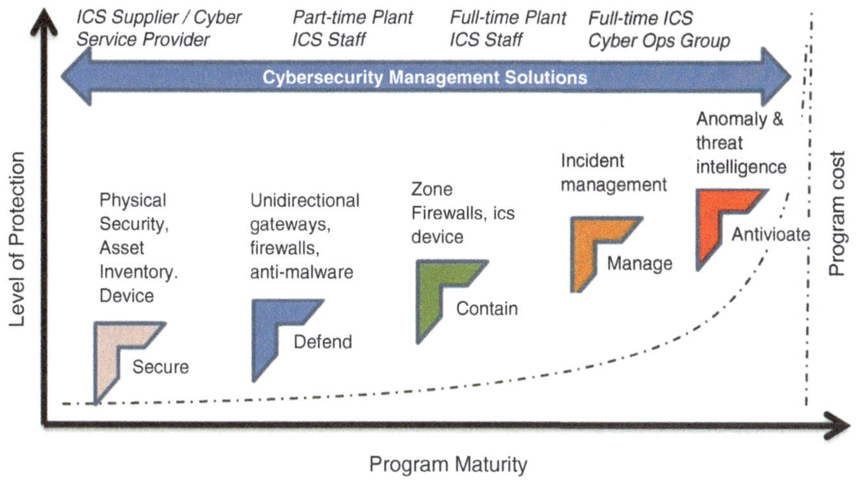

Fig. 11.4 Cybersecurity management system

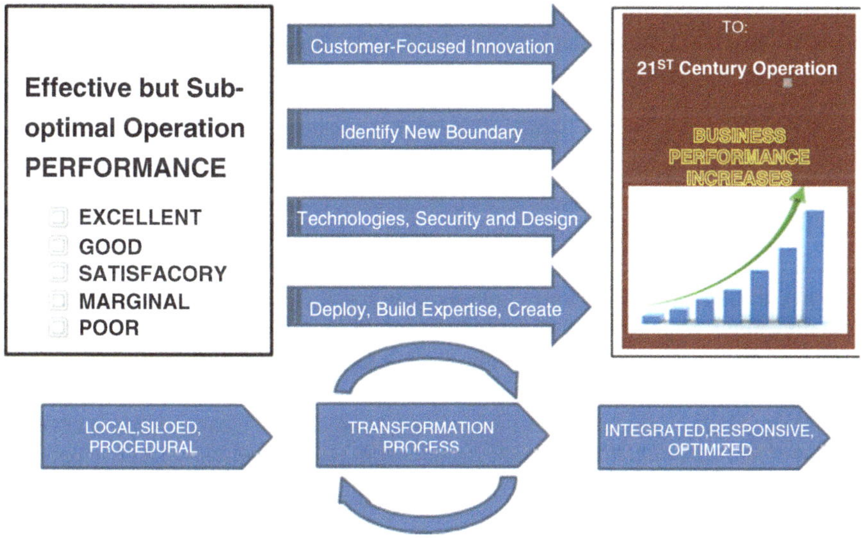

Fig. 11.5 Digital transformation

Protocols in IIoT

The IIoT infrastructure uses client/server *protocols* for transmitting data. These protocols are used to show how the request and response roles are executed and define the technique that is to be monitored in a smart industrial environment.

- *OPC UA*

OPC Unified Architecture (OPC UA) is a machine-to-machine communication protocol used for industrial automation designed and developed by the OPC Foundation. Some distinguished characteristics are as follows:

- Emphasis on connecting with manufacturing tools and data gathering
- Open source and freely accessible and maintainable under GPL 2.0 license
- Cross platform not linked with the operating system or system language
- Service-oriented architecture (SOA)
- Essential complication
- Robust safety

The vendors and organizations model their complex data to a service-oriented architecture of OPC UA protocol based on integral information model for forming statistical infrastructure. This protocol is used by industries such as pharmaceuticals, oil and gas, building automation, industrial robotics, security, manufacturing, and process control for functioning and performing optimistically.

- *HTTP (REST/JSON)*

REST stands for *Re*presentational *S*tate *T*ransfer that depends on a displaced client-server and primary infrastructures. In most of the cases, the transmission is made using the HTTP protocol. JSON (JavaScript Object Notation) is a lightweight data-interchange format that makes easy and user friendly for reading and writing for the users.

- *MQTT [14]*

MQTT (*Message Queuing Telemetry Transport*) is an ISO standard publish-subscribe-based messaging protocol. It works on the upper layer of the TCP/IP protocol. It is measured for connections with distant locations with limited network bandwidth. An MQTT system contains customers connecting with a server frequently called as "broker." A client may be an originator of data or a subscriber. All the clients can get connected with a broker for performing a task.

To access an information, the procedure is arranged in the form of a hierarchy where a publisher can distribute a new data set and control messages are sent to the broker. The brokers distribute the information to the clients who need to access it. The publisher need not have the information about the data subscribers and vice versa. While publishing the client first gets connected with a broker and sends a default message as an acknowledgment to the subscribers if the brokers detect the publishing client (Fig. 11.6).

- *CoAP [6]*

Constrained Application Protocol (CoAP) is a protocol that is used to specify how the low-powered computing constrained devices can be operated in the IoT environment. This protocol was designed by Internet Engineering Task Force (ITEF) for enabling simple devices and controlling them to join IoT by using constrained network with low bandwidth and availability. CoAP functions as

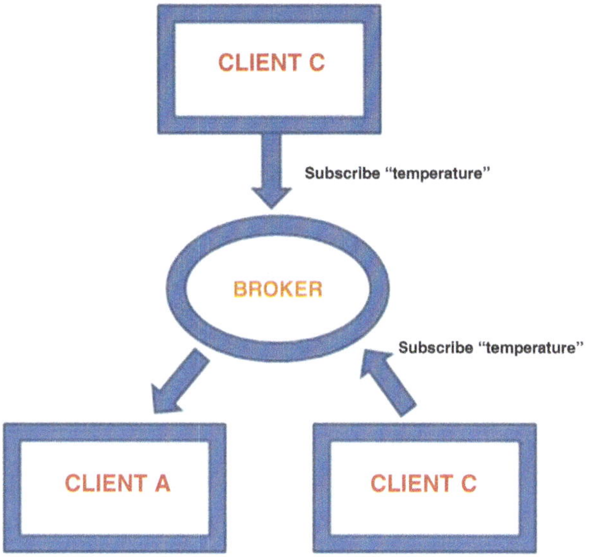

Fig. 11.6 Example for MQTT protocol

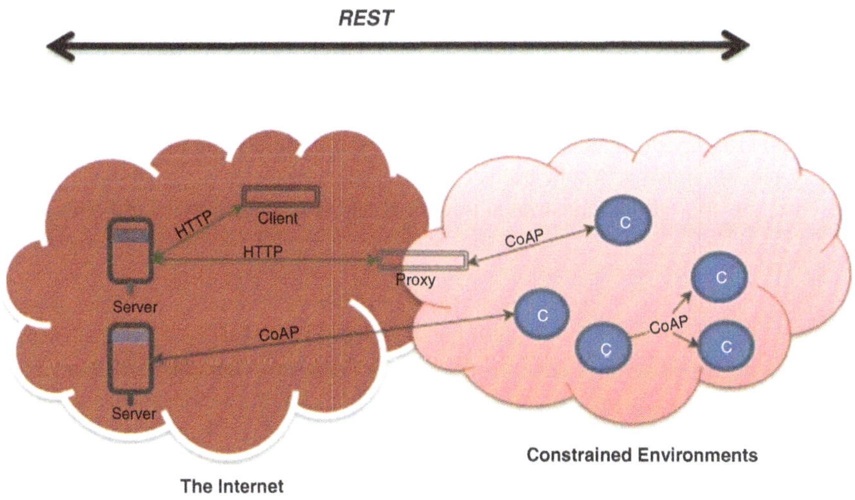

Fig. 11.7 CoAP architecture

aHTTP for constrained devices that perform as a sensor or an actuator used in machine-to-machine (M2M) communication. Finally, this protocol is designed for better reliability with low bandwidth and high congestion with reduced overhead (Fig. 11.7).

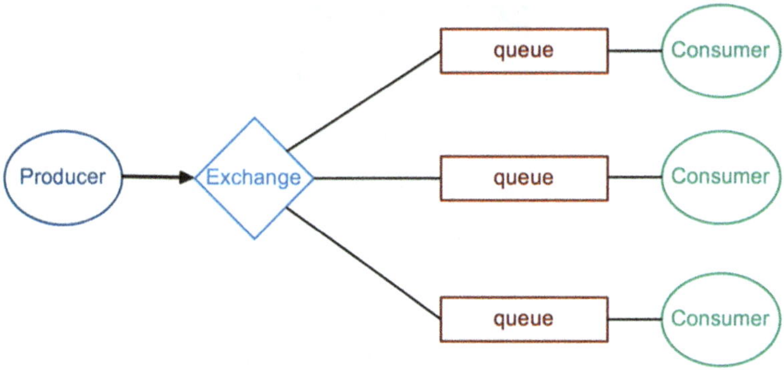

Fig. 11.8 AMQP system

- *DDS:*

The Data Distribution Service (*DDS*) is a networking standard that links the sensor nodes to the cloud server. DDS uses a connectionless data model with the ability to publish and subscribe to data with the desired quality of service (QoS). A DDS-based system does not interact with the application systems. A DataBus exists in the structure that robotically determines and attaches publishing and subscribing applications. The DataBus contests and imposes QoS. DDS disables problems related with system-to-system integration, such as lack of scalability, interoperability, and the ability to evolve the architecture. It allows plug and play simplicity, scalability, and especially high performance. Possibly the best way to understand DDS is to inspect the systems that use it. Applications extend to the areas such as healthcare, energy, defense, transportation, industrial automation, and communications industries.

- *AMQP*

AMQP stands for Asynchronous *Message Queuing Protocol,* which is an open standard that uses binary application layer protocol designed for message-oriented middleware for communicating producers, brokers, and consumers through messages that increase coupling and scalability. The AMQP enables a wide range of submissions and assemblies to work together, irrespective of their inner designs, regulating enterprise messaging on an manufacturing scale (Fig. 11.8).

Conclusion

Thus, the IoT ecosystem has been adopted by the existing industries for process improvements, better quality management, cost–effectiveness, and improved efficiency. The upcoming industries utilize the strength of IoT infrastructure for product optimization and using big data analytics from thousands of tiny sensors for data gathering. Industrial IoT is fast growing technology with limitless possibilities for future industries and manufacturing units. This chapter briefly discusses the innovations, challenges, and protocols used by the IIoT systems.

References

1. Hofmann, E., & Rüsch, M. (2017). Industry 4.0 and the current status as well as future prospects on logistics. *Computers in Industry, 89*, 23–34.
2. Ungurean, I., & Gaitan, V. G. (2014). An IoT architecture for things from industrial environment. In: *Communications (COMM), IEEE 2014 10th International Conference on* (pp. 1–4).
3. Gorecky, D., Schmitt, M., Loskyll, M. et al. (2014). Human-machine interaction in the Industry 4.0 era. In: *The 12th IEEE International Conference on Industrial Informatics (INDIN)*, Porto Alegre, Brazil (pp. 289–294).
4. Lu, Y. (2017). Industry 4.0: A survey on technologies, applications and open research issues. *Journal of Industrial Information Integration, 6*, 1–10. Available at: http://www.sciencedirect.com/science/article/pii/S2452414X17300043. Accessed May 14, 2017.
5. McKinsey Digital. (2015). *Industrie 4.0. How to navigate the digitization of the manufacturing sector, Berlin, Düsseldorf, and Munich*. McKinsey.
6. "Industry 4.0 and manufacturing ecosystems" by Deloitte University Press
7. Gossman, W. E., & Hartmaier, P. J.. (2001). *System and method for providing data to a wireless device upon detection of activity of the device on a wireless network*. US Patent 6317594 B1, 13 Nov 2001.
8. Muhtaroglu, F. C. P., Demir, S., Obalı, M., & Girgin, C.. (2013). Business model canvas perspective on big data applications. In: *Proceedings of the 2013 IEEE International Conference on Big Data*, Santa Clara, CA, USA (pp. 32–37).
9. Lee, E. A. (2008). Cyber physical systems: Design challenges. In: *The 11th IEEE international symposium on object oriented realtime distributed computing (ISORC)*, Orlando, Fl (pp. 363–369).
10. Kang, W., Kapitanova, K., Son, S. H., et al. (2012). RDDS: A real-time data distribution service for cyber-physical systems. *IEEE Transactions on Industrial Electronics, 8*(2), 393–405.
11. Zhang, Y., Xiang, Y., & Huang, X. (2017). A cross-layer key establishment model for wireless devices in cyber-physical systems. In: *Proceedings of the 3rd ACM Workshop on Cyber-Physical System Security, CPSS@AsiaCCS 2017, Abu Dhabi*, 2 Apr 2017 (pp. 43–53).
12. Hwang, J.-J., Chuang, H.-K., Hsu, Y.-C., & Wu, C.-H.. (2011). A business model for cloud computing based on a separate encryption and decryption service. In: *Proceedings of the 2011 International Conference on Information Science and Applications*, Jeju Island, South Korea (pp. 1–7).
13. Carter, C. R., & Ellram, L. M. (2003). Thirty-five years of the journal of supply chain management: Where have we been and where are we going? *The Journal of Supply Chain Management, 39*(2), 27–39.
14. https://info.smartindustry.com/si16-203-lp-kepware

Chapter 12
Smart Meter: A Key Component for Industry 4.0 in Power Sector

M. Kathiresh and A. Mahaboob Subahani

Introduction

Electric power has become indispensable to human survival and progress. Grid is the electricity system that comprises of power generation, power transmission, power distribution, and power consumption. In conventional power grids, electric power is transmitted from a few centralized power generating stations to a large number of distribution centers with electricity users or customers [1]. A smart grid is a new variant of electric power grid under development, which allows unconventional energy flow and bidirectional data flow to create an advanced automatic and decentralized power distribution network. Apart from the efforts to meet the growing demand, automation in the energy distribution is also necessary to enhance people's life standard [2].

There are basically two types of energy meters, namely, electromechanical and electronic meters. The electromechanical induction meter operates by counting the revolutions of an aluminum disc, which is made to rotate at a speed proportional to the power consumed. The number of revolutions is proportional to the energy usage. The voltage coil consumes a small and relatively constant amount of power, typically not greater than 2 W, which is not registered on the meter. The current coil similarly consumes a small amount of power in proportion to the square of the current flowing through it, typically up to a couple of watts at full load, which is registered on the meter [1–3].

The metallic disc is acted upon by two coils. One coil is connected in such a way that it produces a magnetic flux in proportion to the voltage and the other produces a magnetic flux in proportion to the current. The field of the voltage coil is delayed by 90 degrees using a lag coil. This produces eddy currents in the disc, and the effect

M. Kathiresh (✉) · A. M. Subahani
Department of Electrical and Electronics Engineering, PSG College of Technology, Coimbatore, TN, India

© Springer Nature Switzerland AG 2020

G. R. Kanagachidambaresan et al. (eds.), *Internet of Things for Industry 4.0*, EAI/Springer Innovations in Communication and Computing,
https://doi.org/10.1007/978-3-030-32530-5_12

is such that a force is exerted on the disc in proportion to the product of the instantaneous current and voltage. A permanent magnet exerts an opposing force proportional to the speed of rotation of the disc. The equilibrium between these two opposing forces results in the disc rotating at a speed proportional to the power being used. The disc drives a register mechanism that integrates the speed of the disc over time by counting revolutions, much like the odometer in a car, in order to render a measurement of the total energy used over a period of time for a single-phase AC electricity meter. Additional voltage and current coils are required for a three-phase configuration.

Horizontal aluminum rotor disc is visible in center of meter. The aluminum disc is supported by a spindle which has a worm gear that drives the register. The register is a series of dials which record the amount of energy used. The dials may be of the cyclometer type, an odometer-like display that is easy to read where for each dial a single digit is shown through a window in the face of the meter, or of the pointer type where a pointer indicates each digit. With the dial pointer type, adjacent pointers generally rotate in opposite directions due to the gearing mechanism. The amount of energy represented by one revolution of the disc is denoted by the symbol which is given in units of watt-hours per revolution.

Electronic meters display the energy used on an LCD (liquid crystal display) or LED (light-emitting diode) display, and can also transmit readings to remote places. Apart from the amount of electricity used, the electronic electric meter type can also record other parameters of the load and supply such as maximum demand, power factor, and reactive power used. Some of the advanced kinds of electric meters also include electronic clock mechanisms to compute a value, rather than an amount of electricity consumed, with the pricing varying by the time of day, day of week, and seasonally.

The technology used in most of the solid-state electric meter type is the use of a current transformer to measure the current. The main current-carrying conductors need to pass through the meter itself and so the meter can be located remotely from the main current-carrying conductors, which is a particular advantage in large-power installations.

The modules involved in the conventional and smart metering system are shown in Fig. 12.1. Traditional meter reading is inefficient to meet the future residential and industrial developmental needs. Hence, there is an increased demand for automatic meter reading systems, which collect meter readings without any human intervention [3]. Collection of meter reading is also inefficient because a meter reader has to physically be onsite to utility environment. Traditional electromechanical meters used today are prone to drift over temperature. This method of collecting of meter readings becomes more problematic and costly when readings have to be collected from vast and often scattered rural areas. The older electromechanical and present electronic meter readings are taken by persons involved in the job, and the amount is noted down in the electricity board card. This may consume a lot of time and doubtlessly require many human resources. Fraudulent customers could cheat without being detected for several years. The tariff for the energy consumption is

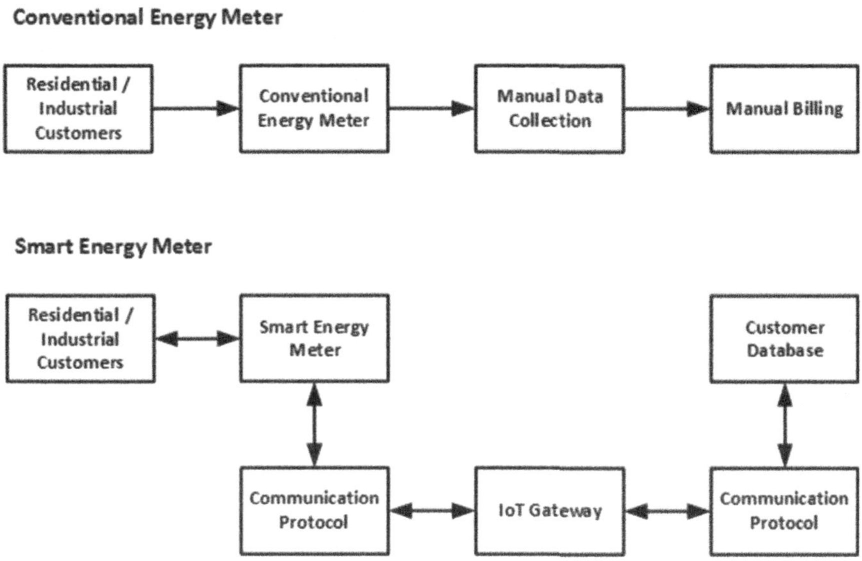

Fig. 12.1 Components of conventional meter and smart meter

calculated based on the readings taken by people from the electricity board. Hence, it is more prone to error.

Smart meters are the next generation of meters, which can replace existing energy meters and offer a range of benefits, both for the individual electricity and gas consumer and for the network systems in general. Through automated recording, actual energy usage over short intervals and automated communicating of metering data to the network operator, the smart meter eliminates the need for a home visit to manually read the meter and facilitate more cost-efficient network operations [4].

Electricity suppliers will be able to offer innovative pricing arrangements for customers to support the efficient use of energy (i.e., time-of-use tariffs). But the smart meter also offers potential advantages directly to customers. Smart meters can be an alternative to the concept of estimated billing system, meaning customers will only be billed for the electricity they actually use and smart meters allow easier switching of energy supplier to get the best deals. However, the most immediate consumer benefit of the smart meter is the potential to receive more frequent and more detailed information on how much energy is being used (also called feedback). For instance, households with a smart meter will be offered to receive consumption and cost statement from the supplier every 2 months, hereinafter referred to as the home energy report. At the specific request of the consumer, the grid operator will forward the smart metering data on a more frequent (e.g., daily) basis to the energy supplier – or another service provider – for additional analysis of the consumer's energy consumption over the previous days, weeks, and months.

For monitoring of real-time energy consumption, customers can add a wireless communication interface, in-home display or online application to their smart meter. Products and services are emerging on the market for this purpose. In order to support product and market development, functional requirements for the communication interface have been established in legislation. The grid operators have also established technical requirements for this purpose.

Smart meters can read real-time energy consumption information including the values of voltage, phase angle, and the frequency and securely communicates that data. The ability of smart meters for bidirectional communication of data facilitates the process of collecting information regarding the electricity fed back to the power grid from customer premises. Smart meters can communicate and execute control commands remotely as well as locally. They can be used to monitor and also to control all home appliances and devices at customer premises.

Smart meters are usually capable of much more than this by including the ability to communicate with in-home appliances, programmable communicating thermostats, and other loads. Smart meters can record consumption data at intervals (as frequently as every minute) and automatically transfer the information to the utility over a secure network [5]. Various communication architectures, including point-to-multipoint and mesh networks, have been implemented. This network, often in conjunction with a backhaul layer, provides two-way connectivity between the utility and the meter. The network also supports pushing signals to the meter, which could be used for an "on demand" reading to confirm power restoration after an outage, or to notify the meter of an upcoming "super peak" event. When the meter is notified of an event, the information is communicated to appliances inside the customer's home using the meter's onboard GSM radio chip or other means. Based on predefined preferences or rules, different appliances can act to reduce or shift consumption in preparation for or during the "super peak" events.

They can measure electricity consumption from the grid, support decentralized generation sources and energy storage devices, and bill the customer accordingly. Smart meters can be programmed such that, only power consumed from the utility grid is billed while the power consumed from the distributed generation sources or storage devices owned by the customers are not billed. Smart meters can limit the maximum electricity consumption, and can terminate or reconnect electricity supply to any customer remotely.

A smart meter system employs several control devices, various sensors to identify parameters and devices to transfer the data and command signals. In future electricity distribution grids, smart meters would play an important role in monitoring the performance and the energy usage characteristics of the load on the grid. Collection of energy consumption data from all customers on a regular basis allows the utility companies to manage electricity demand more efficiently and also to advise the customers about the cost-efficient ways to use their appliances. Smart meters can be programmed to maintain a schedule for operation of the home appliances and control operation of other devices accordingly.

In addition, integration of smart meters helps utility companies in detecting unauthorized consumption and electricity theft in view of improving the distribution efficiency and power quality. Design of future electricity markets are aimed at providing their customers with highly reliable, flexible, readily accessible, and cost-effective energy services by exploiting advantages of both large centralized generators and small distributed power generation devices. In addition, distributed generation would be an essential integral part of future household energy systems. Utility companies try to identify more profitable customers to provide optional value-added services, as smart meters can identify such customers based on the distributed generation sources and overall power consumption.

Smart Metering for Developing Countries

In many developing countries, conventional energy meters are used for billing the energy consumed by customers. For ease of operation of the home appliances, monitoring the grid, improving the power quality, improved load sharing, detecting non-technical losses, and other implied advantages, smart meters are to be introduced in developing countries. Power utility companies worldwide lose about 20 billion dollars each year because of nontechnical losses. In addition, growing nontechnical losses because of theft and billing irregularities force the utility companies to implement a transparent and genuine metering system. However, deployment of smart grid and smart meter system involves huge budgets. It would be very difficult for utility companies to invest billions of dollars on an infrastructural upgrade that has no direct return on the investment. Hence, smart meters with minimum required, but essential features may be designed for implementation in countries with weaker economy. Moreover, smart meters might not be implemented for luxury in operation, but they must be introduced in order to fight the basic problems that power utility companies and its customers face [6, 7].

Smart meters with great networking capability and advanced software tools are difficult to tamper and hack, which improves the distribution efficiency. Integration of smart meters enhances facilitation of decentralized generation and power storage devices. In the near future, total energy demand is expected to become double the current electricity demand. In view of this situation, many developing countries do not have resources for the additional capacity addition. To fill this gap, apart from increasing the installed generation capacity, controlling the electricity theft and regularizing the existing electricity customers can manage the load within demands.

Benefits of Smart Meters over Conventional Meters

The smart meter installation brings quite evident benefits to the utilities in a medium or short span of time. It provides a more timely and precise billing and reduces the costs of interacting with customers by activating, closing, or suspending contracts with no personnel displacement. It also offers significant help in avoiding electricity-related frauds, and finally it offers the potential for better power peak control and distribution. The interest of electrical energy distributors is therefore self-evident, and it is reasonable to assume that, even for other energy sources (gas, water, and heat), the utilities could find it technologically sound and profitable to extend a similar smart metering philosophy, even though some delay can be expected due to the lack of such a preexisting physical interconnection as it is the low-voltage power line for the electricity meter [8].

On the other hand, alternatives (e.g., wireless solutions) are available, and regulatory authority decisions can offer additional obligations or incentives. Nonetheless, a complete exploitation of smart metering advantages is only achieved when an efficient local interface between the smart meter and the household is accomplished. Benefits can then be extended to the community and to the final user. As a matter of fact, the interconnection of energy users and the ability to easily provide data related to energy consumption are crucial opportunities for the management of such a critical area as the energy saving. After the Kyoto Protocol, even the most recent 20-20-20 agreements converge at implementing policies aimed at saving energy and reducing CO_2 emissions.

Design of a Smart Meter Prototype

In the smart meter, potential transformer (PT) and current transformer (CT) are used for measuring current and voltage respectively. The sensed sinusoidal signal from CT and PT is given to the signal conditioning unit where the signal is rectified or clamped to a reference voltage of half of its peak magnitude in order to shift the negative half into positive side. The rectified signal is then given to a microcontroller for processing. The microcontroller calculates the power consumption using the voltage and current signals, and the power factor, which is obtained by sampling the magnitude of the voltage and current.

Energy is calculated by counting the power consumption time using the inbuilt timer and multiplying it with the power consumption. Apart from this, prescribed peak load time from the consumer will be noted and alarm will be given to the consumer. During the peak hours, consumers can turn ON or turn OFF the loads remotely through the Internet Gateway module (ESP8266). Figure 12.2 shows the various functional blocks of the smart meter system.

Current Measurement

The circuit shown in Fig. 12.3 is used for current measurement, which measures the total current drawn from the supply. The secondary side of CT produces current which is reduced and proportionally varying to the primary current to be measured. This current is converted to voltage with the help of a shunt resistor called *burden*

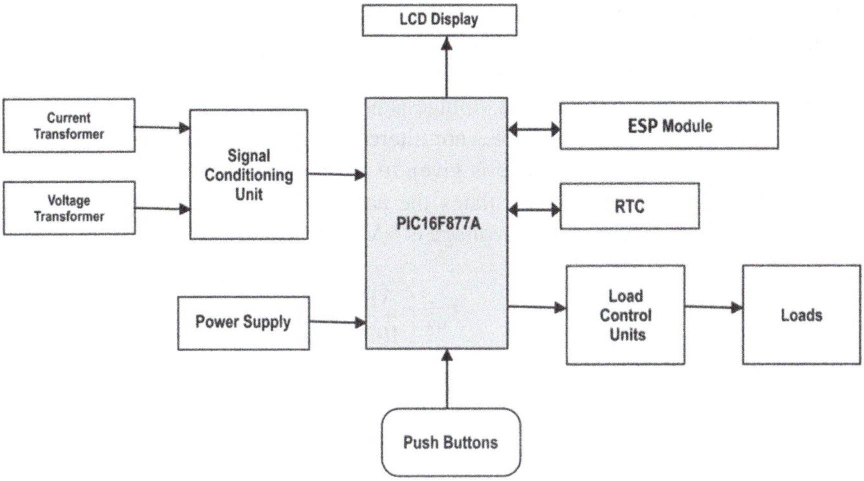

Fig. 12.2 Structure of the smart meter prototype

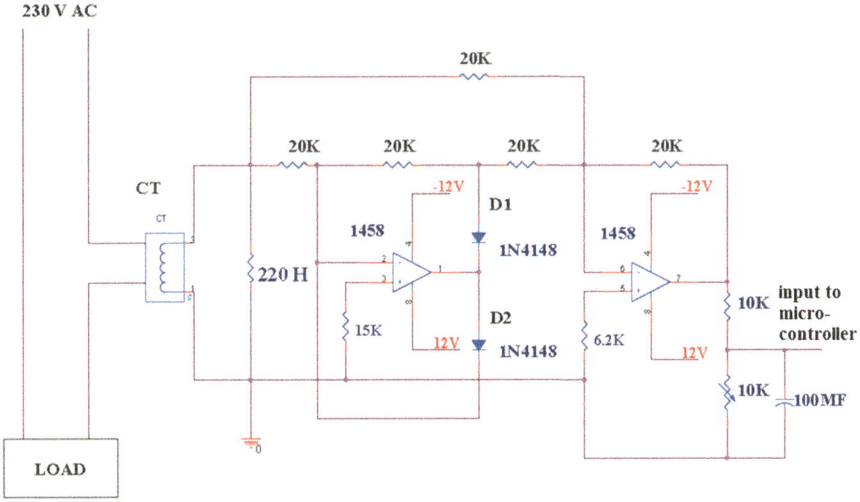

Fig. 12.3 Current measurement circuit

resister. The shunt resister voltage is rectified by an Op-Amp precision rectifier in order to have a circuit behaving like an ideal diode or rectifier. The Op-Amp precision rectifier is a combination of a half-wave precision rectifier and a summing amplifier. It operates by producing an inverted half-wave-rectified signal and then adding that signal to the original signal in the summing amplifier.

When the input is positive, the diode D1 is ON and D2 is OFF, so both the op-amp1 and op-amp2 act as an inverter. Hence the output voltage is equal to the input. When the input is negative, the diode D1 is OFF and D2 is ON, and the differential input to the op-amp2 is zero. Hence, the output voltage is equaled positive of the input voltage. This circuit is also known as absolute value circuit as the output is always positive for both positive input and negative input.

Then the output of the rectified voltage is adjusted to 0–5 V with the help of a variable resistor. Then, output ripples are filtered by the capacitor connected across the output. The output DC voltage is given to the microcontroller for further processing. The microcontroller calculates the actual current value using Eq. 12.1. (assuming that the ADC reference voltage is 5 V and the output resolution is 10 bit).

$$I_{\text{rms}} = \frac{I_{\text{ADC}}\left(\frac{311}{1023}\right)}{\sqrt{2}} \tag{12.1}$$

Voltage Measurement

Figure 12.4 shows the circuit diagram for voltage measurement which measures the supply voltage. The secondary side of PT produces voltage which is reduced and proportionally varying to the primary voltage to be measured. This voltage is rectified and filtered by the same Op-Amp precision rectifier circuit that was discussed in the above section.

Then the output of the rectified voltage is adjusted to 0–5 V with the help of a variable resistor. Subsequently, output ripples are filtered by the capacitor connected across the output. The output DC voltage is given to the microcontroller for further

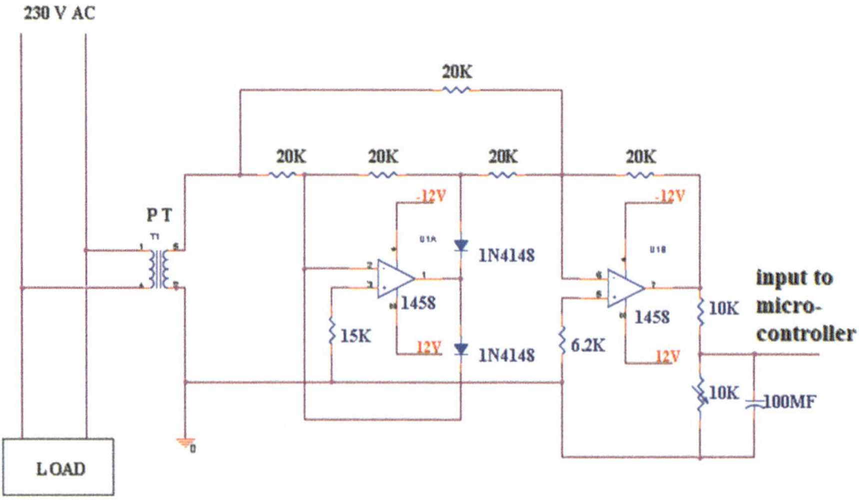

Fig. 12.4 Voltage measurement circuit

processing. The microcontroller calculates the actual current value using Eq. (12.2) (assuming that the ADC reference voltage is 5 V and the output resolution is 10 bit).

$$V_{\mathrm{rms}} = \frac{V_{\mathrm{ADC}}\left(\frac{311}{1023}\right)}{\sqrt{2}} \tag{12.2}$$

Power Factor Calculation

Essentially, power factor is a measurement of how effectively electrical power is being utilized in a power system network. The higher the power factor, the more effectively electrical power is being utilized. It is the relationship between active power and the total power consumed or apparent power.

The power factor is also defined as the cosine of the phase angle difference between the voltage and current waveforms. As resistive loads do not produce any phasor difference between the voltage and current, all the useful power is delivered directly and converted to heat. Hence, the power consumed by a resistive load is the real power which is fundamentally the circuit's average power, so the power factor is unity [9].

On the contrary, the reactive loads such as inductor and capacitor produce a phasor difference between the voltage and current. So the power factor will not be unity. In addition, the utility loads are not pure resistive loads at any time. Hence before the power measurement, it is very much needed to calculate the power factor [9, 10].

The idea of calculating power factor is to count the time difference (T_d) between the zero crossing instant of both the voltage (T_v) and current (T_i), and convert the time into degree (θ_d) using the Eqs. (12.3) and (12.4), respectively. This is the phasor difference between the voltage and current, taking the cosine value of it will give the power factor as given in Eq. (12.5).

$$T_{\mathrm{d}} - T_{\mathrm{v}} - T_{\mathrm{i}} \tag{12.3}$$

$$\theta_{\mathrm{d}} = 360^{\circ} f T_{\mathrm{d}} \tag{12.4}$$

Where f is power supply frequency

$$\mathrm{PF} = \cos\theta_{\mathrm{d}} \tag{12.5}$$

Active Power and Energy Calculation

The active power is calculated in Watt, by multiplying the voltage, current, and power factor that is given in Eq. (12.6). Active power measurement is made for every short interval, and they are summed for a period of time as given in Eq. (12.7).

$$P_i = V_{\text{rms},i} I_{\text{rms},i} \text{PF}_i \tag{12.6}$$

$$P_{i+1} = P_i + P_{i+1} \tag{12.7}$$

Once the active power is calculated for a period of time, the energy is calculated in kWh using Eq. (12.8), and the total energy consumption is calculated by the cumulative sum of energy consumption over a period of time using the Eq. (12.9). The total energy consumption in kWh can be used for calculating the number of units consumed and in turn the tariff if the per unit cost is known.

$$E_{p,i} = \frac{P_i \times \text{time}}{1000 \times 3600} \tag{12.8}$$

$$E_{p,i+1} = E_{p,i} + E_{p,i+1} \tag{12.9}$$

Reactive Power and Energy Calculation

The reactive power is calculated in VAR, by multiplying the voltage, current, and sine of the phasor difference between them as given in Eq. (12.10). Reactive power measurement is done for every short interval, and they are summed for a period of time as given in Eq. (12.11).

$$Q_i = V_{\text{rms},i} I_{\text{rms},i} \sin\theta_d \tag{12.10}$$

$$Q_{i+1} = Q_i + Q_{i+1} \tag{12.11}$$

Once the reactive power is calculated for a period of time, the energy is calculated in VARh using Eq. (12.12), and the total energy consumption is calculated by the cumulative sum of energy consumption over a period of time using the Eq. (12.13). The total reactive energy consumption in VARh can be used for calculating the penalty for industrial customers if it exceeds the limit.

$$E_{p,i} = \frac{Q_i \times \text{time}}{3600} \tag{12.12}$$

$$E_{Q,i+1} = E_{Q,i} + E_{Q,i+1} \tag{12.13}$$

Real-Time Clock (RTC)

RTCs are designed for ultra-low power consumption as they usually continue running when the main system is powered down. The low power DS3231 RTC module is used as real-time clock which can work either in 24-hour mode or 12-hour mode with AM/PM indicator. It automatically adjusts for months fewer than 31 days including leap year compensation up to year 2100. It also comes with built-in power sensing circuit which senses power failures and automatically switches to back up supply of 3 V CMOS Battery. The address and data are transferred from RTC to the microcontroller serially through an I^2C, bidirectional bus.

Load Control Unit

A relay is a type of actuator that provides connection between two terminals or points in association with a control signal. In simple terms, a relay is an electrical switch in which a small electrical current is required to control the switching operation (ON and OFF). The relays, depending on the principle of operation, can be classified as electromechanical relays and solid-state relays. Electromechanical relay uses an electromagnet to achieve the mechanical action of the switch. A small current given to the coil wound on an iron core will energize the coil, and the contacts of the relay move to on position. When the coil is de-energized, the contacts move back to off position.

In contrast to electromechanical relays, which consist of moving parts and magnetic flux, a solid-state relay or SSR does not consist of any moving parts. A solid-state relay consists of semiconductor devices such as transistors, thyristors, SCRs, or TRIACS to achieve the switching operation. So in the prototype developed, 800 V, 10 A, rated SFS2004/59 TRIAC is used for connecting and disconnecting the lamp load with the power supply.

Working of the Smart Meter Prototype

The smart meter prototype consists of potential and current transformers, PIC19F877A microcontroller, IoT gateway (ESP8266), relay control unit, signal conditioning unit, RTC module, and LCD display. Once the smart meter is powered up, it initializes its modules. The rms value of voltage and current is calculated by reading the values from PT and CT, respectively. The power factor is calculated by measuring the phase difference between the voltage and current. Then the instantaneous power is calculated by multiplying all the parameters that are calculated. This results in the instantaneous power consumption of the load. This process is repeated for few minutes and the cumulative power is calculated in watts. Total energy consumption is calculated from the total power and the time till which the power is calculated in kWh. The energy consumption data is sent to smart grid server through an IoT gateway which can help to monitor the total power consumption or demand in any particular area/region. The tariff is calculated by multiplying it with the per-unit cost at the smart grid server, and the same is sent to the customers' mobile phone.

The parameters such as voltage, current, power consumption, and total tariff are displayed in the LCD display. The same is sent to the customers' mobile through SMS on request. The workflow of the prototype is shown in Fig. 12.5. The load demand data from the server is fed to the smart meter continuously to enable the meter for optimizing the load pattern of the customer. With the help of real-time clock, the peak hours at which the power consumption from the grid is high is identified and the customer is given alert about the start time of peak load. Hence, the

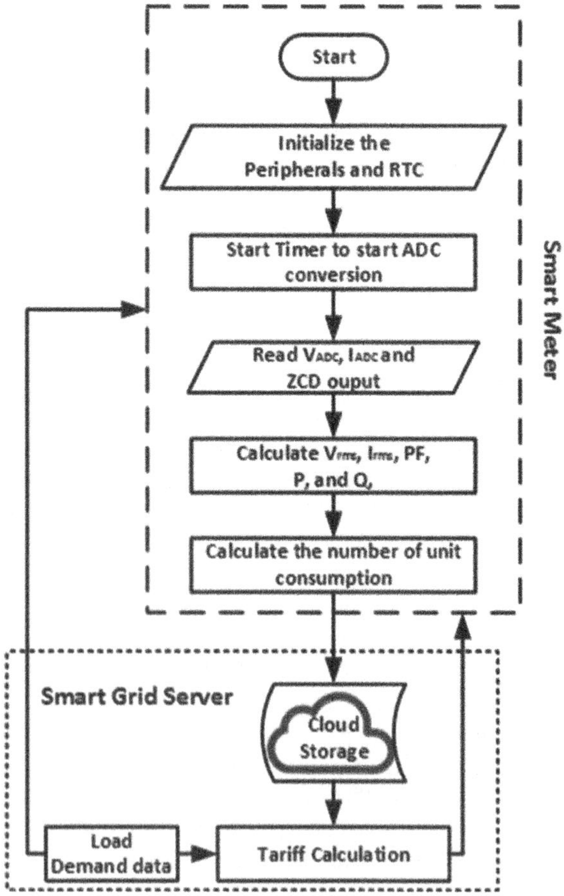

Fig. 12.5 Workflow of the smart meter prototype

customer can turn OFF the loads based on his need which will save him from higher tariff rate. In addition, suggestions are given to the customers for task scheduling for optimum usage of energy to save tariff.

Simulation Results

Using Proteus 8 simulator, the prototype model of smart meter is simulated. In the simulation, PIC microcontroller is used which is interfaced with voltage and current measurement unit, signal conditioning unit, and load control devices. All the parameters such as voltage, current, power factor, and power consumption data are displayed in 20x4 LCD as shown in Fig. 12.6.

As mentioned in the above section, the prototype developed is given with the total load demand data at periodic intervals, by which it identifies the peak hour and alerts the customer. The same is displayed in the LCD as shown in Fig. 12.7a. Once the peak hour is identified, the smart meter starts optimizing the loads that are currently connected to the grid. During this optimization, power consumption of the daily routine task of the customer is also taken into account along with the individual load demand data. The smart meter then prioritizes the tasks based on all the abovementioned parameters and the suggestions are given to the customer to reduce the load either by turning OFF any load or by rescheduling the task. In the simulation results, the same condition is also verified during the peak hour, and the suggestion made by the smart meter is highlighted in Fig. 12.7b. For simulation purpose, random data are used for load demand and the power consumption of daily routine tasks.

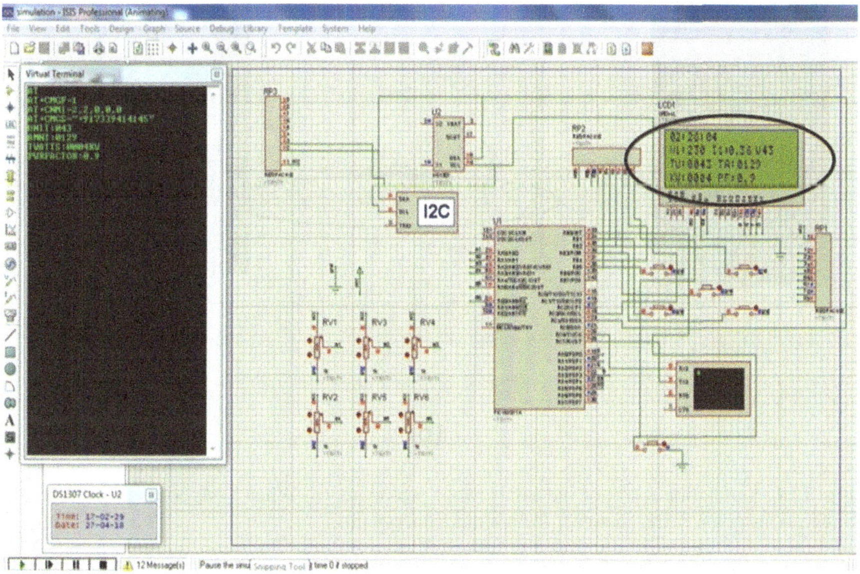

Fig. 12.6 Simulation result of the smart meter prototype

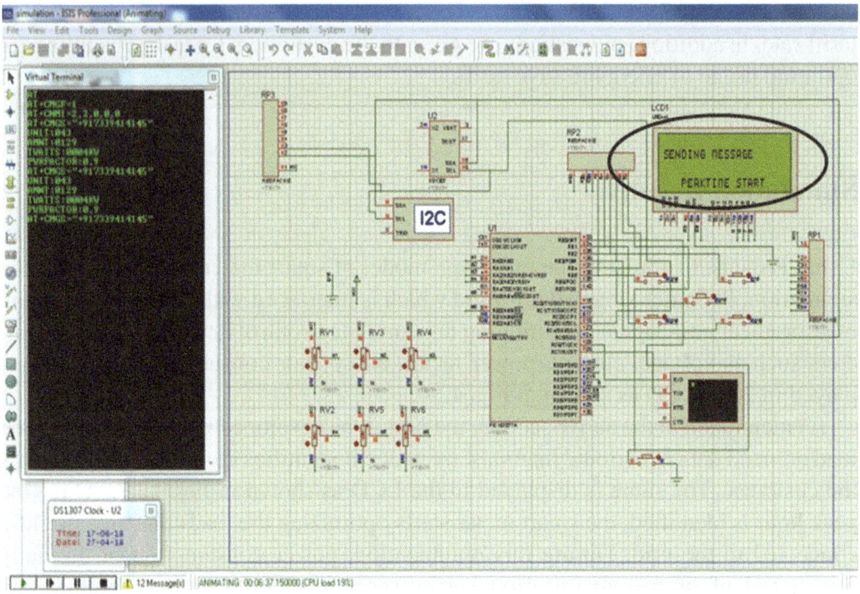

(a)

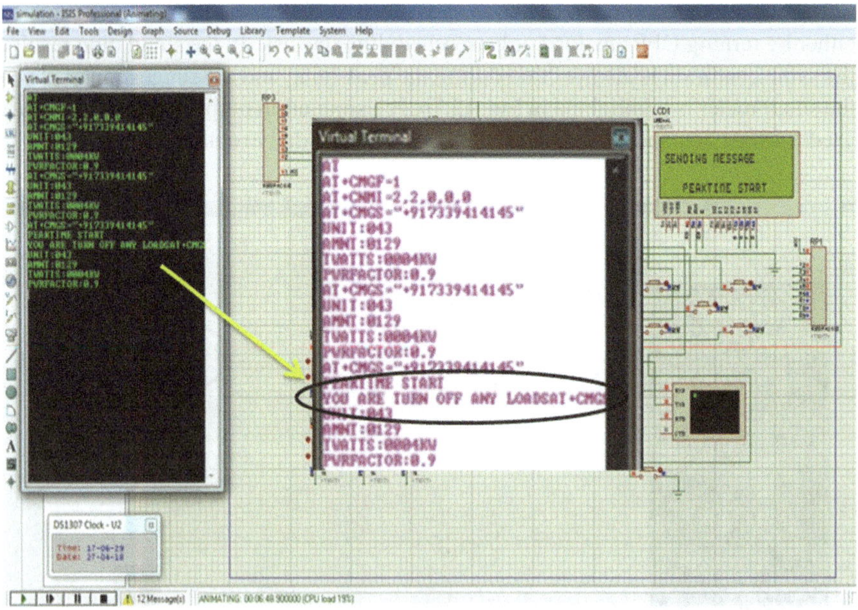

(b)

Fig. 12.7 Simulation result of the smart meter prototype. (**a**) When peak hour is identified. (**b**) Suggestions given to the customer for optimum energy use

Hardware Results

The prototype of smart meter is developed in the laboratory which is shown in Fig. 12.8. It consists of power supply module, measuring unit (PT and CT), zero crossing detector (ZCD), signal conditioning unit, microcontroller, real-time clock, IoT gateway, and LCD display.

For experimental purpose, lamp loads are used, and TRIAC is used for load control. The power consumption and the other parameters are displayed in LCD as shown in Fig. 12.9a. The load demand data is given to the prototype to verify the response in peak hour. The smart meter identified the peak load demand and displayed the warning command in the LCD which is shown in Fig. 12.9b. Further, the power consumption and tariff details are sent to the customers' mobile as SMS along with the peak hour alert as shown in Fig. 12.10.

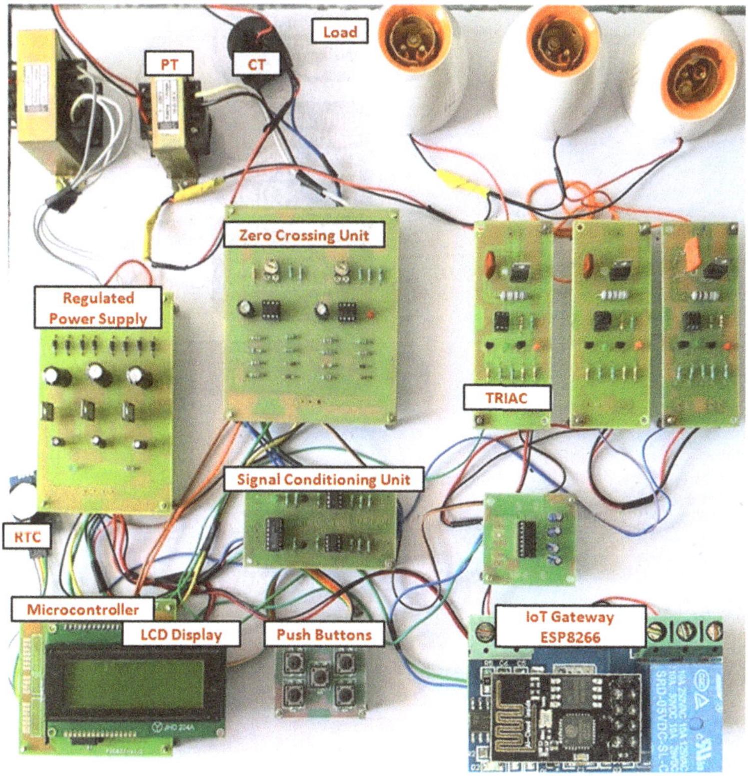

Fig. 12.8 Experimental setup of the smart meter prototype

(a) (b)

Fig. 12.9 Display command of the smart meter prototype. (**a**) Power consumption. (**b**) Peak our alert

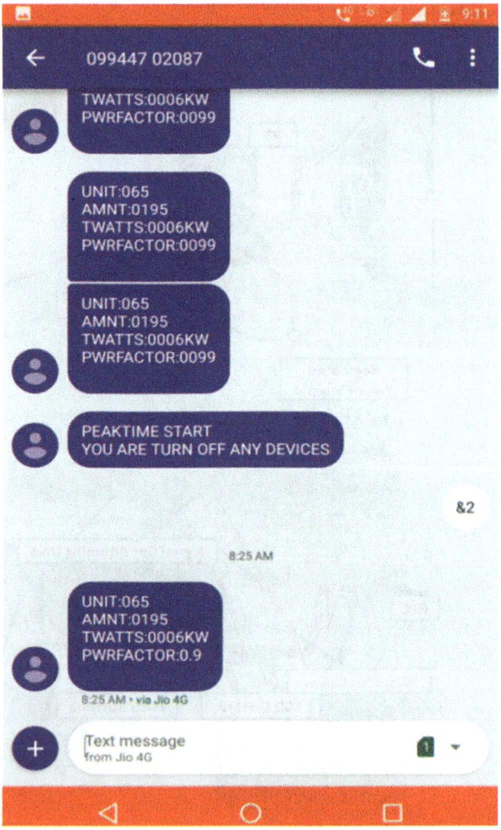

Fig. 12.10 Intimation to the customers' mobile

Implementation of Smart Meter with Prepaid Facility

An attempt is made to develop a smart meter with prepaid facility like the facilities available with mobile networks and direct-to-home (DTH) services. In the prepaid smart meter model, the customer has to recharge the balance by purchasing the recharge cards. The smart meter calculates the total energy consumption and the equivalent tariff for every 1 minute, and the tariff amount is debited from the balance. In this scheme, a minimum balance is fixed, and as long as the balance is greater than the minimum balance, the process continues. If the balance is less than the minimum balance, the customer is given alert and prompted to recharge. If the balance becomes zero, then a warning message is sent to the customer giving him a grace period of 30 minutes to recharge. If the meter is not recharged even after the grace period, the supply is disconnected for the customer with an intimation message. Once this happened, the customer has to contact the energy provider for fixing the issue. He will be given power supply with a penalty that is derived by the supply provider. Algorithm for the prepaid smart meter is given below.

Algorithm for prepaid smart meter
Start
Step 1: Clear all registers
Step 2: Initialize Timers *T0* (interval in min.), *T1*
Step 3: Data acquisition

Voltage conversion:
Start Conversion
Read Digital value and convert it to its equivalent analog value
Find the sum of the squares of the voltage samples $\sum_1^n v_n^2$
Increase the count '*n*'

Current conversion:
Start Conversion
Read Digital value and convert it to its equivalent analog value
Find the sum of the squares of the current samples $\sum_1^n i_n^2$
Increase the count '*m*'

Go to: Step 3

Step 4: Detect positive zero crossing of Voltage:
Start timer *T1*
Find Voltage rms ($V_{rms}=\sqrt{\sum_1^n v_n^2 / n}$)

Increase cycle count 'c_v'
Return

Step 5: Detect positive zero crossing of Current:
Stop timer $T1$
Calculate phase angle difference and power factor ($\theta = \frac{360^\circ}{0.02} \times t_d$; PF = cos θ)

Calculate current rms ($I_{rms} = \sqrt{\sum_1^n i_n^2 / n}$)
Increase cycle count 'c_i'

Step 6: If timer T0 = 0
Calculate the following
Average voltage rms=V_{rms}/c_v
Average current rms=I_{rms}/c_i
Average PF = cos θ/c_i
Real Power = $V_{rms}I_{rms}$ cos θ in kW
Number of units is equal to [Real Power × Time period (interval)]
Calculate the tariff from the number of units
Display all the values calculated above

Step 7: Check the available balance if > tariff
Check the available balance if < Rs. 10/-
Send the customer a warning message to recharge
Subtract the tariff from the prepaid balance

Go to: Step 1

Step 8: If the balance < tariff
Intimate the customer that the supply will be cut off in 5 minutes
Set the timer (T1) to 30 min.
Calculate the cumulative tariff

Step 7: If the timer T1 = 0
Check if the balance < tariff
Turn off the supply

Stop

Else go to: step 7

The flowchart of the smart meter with prepaid facility is shown in Fig. 12.11. The customer is supplied with energy till the available balance is greater than the tariff amount calculated. If not, the smart meter will execute the subroutine A till the grace period. Here the grace time is set to 30 min. If the customer recharges the balance within the grace period, then the smart meter does nothing, but continues to supply energy to the customer by entering into the point B. But if he does not recharge the balance, the customer is disconnected from the grid immediately and intimation is sent to the customer.

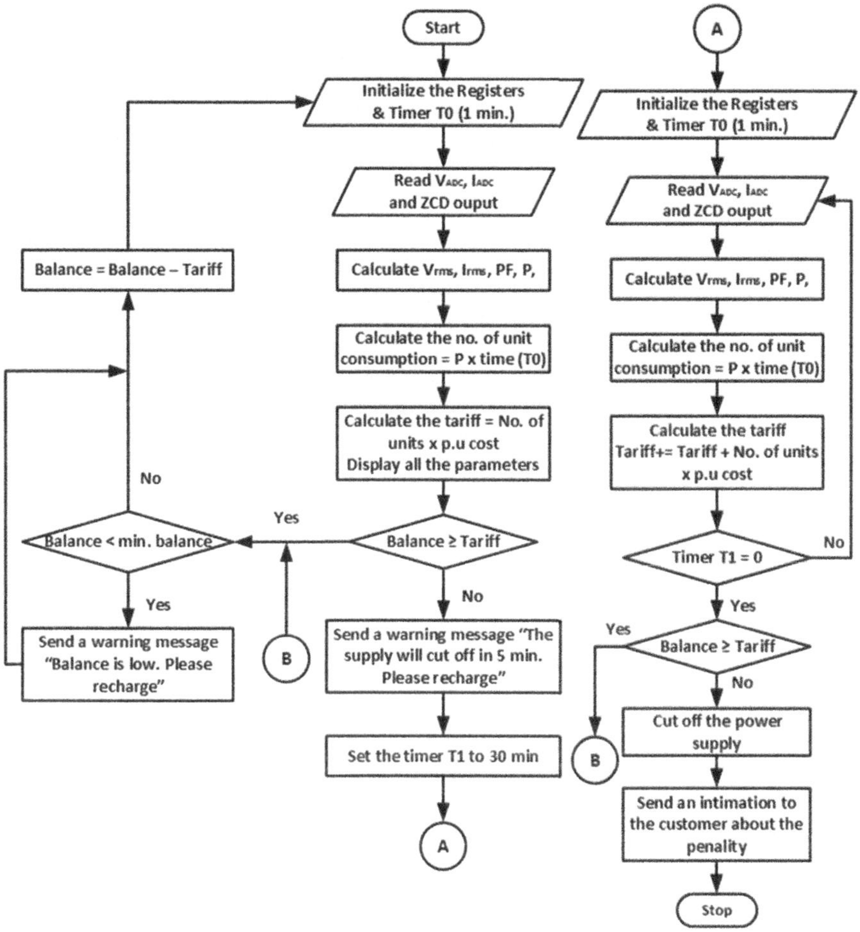

Fig. 12.11 Flowchart of the smart meter with prepaid facility

Smart Meter Applications

The function of smart meter can be further extended to the following applications such as outage detection (OD), Automated Control of Home Appliances (ACHA), demand response (DR), distributed generation systems (DGs), and electronic vehicle charging (EVC).

References

1. Gao, C., & Redfern, M. A. (2011, September 5–8). A review of voltage control in smart grid and smart metering technologies on distribution networks. In *Proceedings 46th International Universities' Power Engineering Conference*, Soest.
2. Barai, G. R., Krishnan, S., & Venkatesh, B. (2015). Smart metering and functionalities of smart meters in smart grid – A review. In *2015 IEEE Electrical Power and Energy Conference (EPEC)*, London, ON (pp. 138–145). https://doi.org/10.1109/EPEC.2015.7379940.
3. Bansal, P., & Singh, A. (2016). Smart metering in smart grid framework: A review. In *2016 Fourth International Conference on Parallel, Distributed and Grid Computing (PDGC)*, Waknaghat (pp. 174–176). https://doi.org/10.1109/PDGC.2016.7913139.
4. Barman, B. K., Yadav, S. N., Kumar, S., & Gope, S. (2018). IOT based smart energy meter for efficient energy utilization in smart grid. In *2018 2nd International Conference on Power, Energy and Environment: Towards Smart Technology (ICEPE)*, Shillong, India (pp. 1–5).
5. Luan, W., Peng, J., Maras, M., Lo, J., & Harapnuk, B. (2015). Smart meter data analytics for distribution network connectivity verification. *IEEE Transactions on Smart Grid, 6*(4), 1964–1971. https://doi.org/10.1109/TSG.2015.2421304.
6. Al-Ali, A. R., Landolsi, T., Hassan, M. H., Ezzeddine, M., Abdelsalam, M., & Baseet, M. (2018). An IoT-based smart utility meter. In *2018 2nd International Conference on Smart Grid and Smart Cities (ICSGSC)*, Kuala Lumpur (pp. 80–83). https://doi.org/10.1109/EPETSG.2018.8658501.
7. Hafez, O. (2017). Time of use prices considering smart meters and their future implementation in Saudi Arabia smart grid. In *2017 Saudi Arabia Smart Grid (SASG)*, Jeddah (pp. 1–5). https://doi.org/10.1109/SASG.2017.8356504, doi: https://doi.org/10.1109/ICSGSC.2018.8541314
8. Morello, R., De Capua, C., Fulco, G., & Mukhopadhyay, S. C. (2017). A smart power meter to monitor energy flow in smart grids: The role of advanced sensing and IoT in the electric grid of the future. *IEEE Sensors Journal, 17*(23), 7828–7837. https://doi.org/10.1109/JSEN.2017.2760014.
9. Emanuel, A. E. (1993). Apparent and reactive powers in three-phase systems. *European Transactions on Electrical Power, 3*, 7–14.
10. Haben, S., Singleton, C., & Grindrod, P. (2016). Analysis and clustering of residential customers energy behavioral demand using smart meter data. *IEEE Transactions on Smart Grid, 7*(1), 136–144. https://doi.org/10.1109/TSG.2015.2409786.

Chapter 13
Role of Software-Defined Network in Industry 4.0

Vakaavinash Reddy and K. Venkatesh

Introduction

Dueto the innovative advances in the industry, the present IoT gadgets are progressively tending to convert smart by furnishing them with a facility to connect the internet to accumulate real-time data. This advancement has created a revolution in most of the verticals in the industry, for example, Agriculture, Health Care, Manufacturing, Energy and Minerals, and furthermore Automobile by empowering them to equip IoT sensors for easy access information from any place on the planet. The development of Cloud Computing made it easy to coordinate them on a solitary common shared platform not only to log the information but also to control these gadgets remotely by providing Over the Air (OTA) software package updates. These smart devices can be incorporated and controlled utilizing smartphones as well and it was anticipated that the total number of associated IoT gadgets hits up to 25 billion by 2020 [1]. As the number of associated gadgets is extending progressively and becoming smart, the current underlying network infrastructure should also be improved to dodge delay in data packet transmission and make the network even smarter by knowing the importance of data carried by each packet. The priority for critical data packets must be recognized ahead and routed so that the Quality of Service (QoS) will not be undermined for critical IoT operations.

In an Internet of Things (IoT) system, the connected end devices are named as 'motes'. These motes accumulate the information from the encompassing environment by the means transducers and sensors that are associated and transmit the information to the cloud platform or some other feedback mechanisms with the goal

V. Reddy (✉) · K. Venkatesh
Department of Information Technology, SRM IST, Kattankulatur, Chennai, India
e-mail: venkatesh.k@ktr.srmuniv.ac.in

© Springer Nature Switzerland AG 2020

G. R. Kanagachidambaresan et al. (eds.), *Internet of Things for Industry 4.0*,
EAI/Springer Innovations in Communication and Computing,
https://doi.org/10.1007/978-3-030-32530-5_13

that the response can be passed to the servo motors and actuators to control the overall system. These motes are further associated with the border router for sending and receiving information over the internet. Most of the organizations that provide cloud service, such as Google, Microsoft, and Amazon, have propelled an innovative service called IoT as a service that allows clients to associate and work with their motes on top of the online Cloud platform [2]. These service providers have additionally presented their IoT Development boards with inbuilt sensors and Artificial Intelligence (AI) gadgets like Google Home, Echo, and Alexa, which empower the users to direct their gadgets through voice commands and speech, which apparently indicates that the future revolution in technology is going to occur in the IoT.

Software-Defined Networking (SDN) application development in IoT is an approach to manage wireless sensor networks and distributed computing, which supports the administration and enables automatic effective framework designing practices to ensure the overall aim to optimize performance and monitoring [3]. Incorporating SDN is expected to deliver the way in which the static design of traditional frameworks is decentralized and complex while existing systems need all the simpler troubleshooting, monitoring, and versatility. SDN introduces a unified framework insight on a central system by decoupling the routing method (control plane) of the system from the packet forwarding process (data plane) and uniting it with IoT-enabled devices [4]. The control plane includes a minimum of one controller which operates as a brain for SDN network where the entire intelligence is united. Yet at the same time, the centralization has its own specific drawbacks concerning security and elasticity.

This chapter mostly focuses on complications that ascend in IoT while forwarding critical traffic and probabilities to include Software-Defined networking in IoT. It also explains device configuration in the network to acquire appropriate topology and routing table in the SDN controller. The process to incorporate one of the sensors like Temperature, LDR, PIR – Motion, and Humidity sensors into motes will also be discussed. The information that was gathered will be stored in the cloud platform or passed to the feedback system to regulate the encompassing environment of the motes, i.e., for example, if there occurs a fire accident near environment of the mote, then it generates critical data packets that are given higher preference than the additional motes that are producing less prior traffic so that the condition can be controlled and neutralized with least delay.

The essence of this chapter can be addressed as follows. Section "Technologies Incorporated" presents concise information on the hardware equipment and software that are incorporated. Section "Related Works" gives a brief on how SDN application implementation visualizes. The foremost problem is represented in Section "The Problem Definition". Section "Architecture and Test Environment" gives information on the architecture for applying SDN Technology in IoT. The intended solution is explained in Section "The Proposed Solution". The obtained results for the proposed solution are highlighted in Section "Result and Discussion". The conclusion followed by the future work is shared in Section "Conclusion".

Technologies Incorporated

To include Software-Defined Networking (SDN) in the Internet of Things (IoT) there are a few open-source hardware and software requirements that are claimed for information collection from sensors and to route the data packets to the server running on top of the cloud platform. The facts about those components are discussed below.

Raspberry Pi Model 3

Raspberry Pi v3 is a low-cost pocket-sized portable mini-computer that runs with 5 V micro USB power supply. It can run the Linux-based Debian operating system in which the hardware comes with 1GB of RAM and a quad-core processor of 1.2Ghz with built-in wireless network adapter that can be utilized in both client mode and Access Point mode [5]. It can be used in any type of embedded application system wirelessly. Raspberry Pi comes with multiple versions, but version 3 series or further is recommended for installing OpenWRT base ovs package.

OpenWRT OVS

OpenWRT is an open-source project, it is a customizable router firmware based on Linux kernel that can be easily deployed on any Linux system platform and it provides a variety of application-specific bundles. Open vSwitch (OVS) is a package inside the Unified Configuration Interface (UCI) that can be downloaded and installed through CLI or GUI [6]. A Raspberry Pi 3 SD memory is flashed with OpenWRT firmware obtained from OpenWRT compatible device downloads page on the website and accessed GUI through LAN default gateway IP address in the device associated machines web browser. As Raspberry Pi 3 comes with an inbuilt wireless network adapter, it can be configured either in client mode or Access Point mode. Initially, to download the required packages it will be configured in client mode and later changed to AP mode to allow IoT devices to connect with the internet. In the case of Raspberry Pi-2 model B, a USB wireless network adapter is inserted in any one of the available USB ports and installed in the adapter-specific wireless-supported drivers from the UCI package repository (Fig. 13.1).

To perform SDN operations, the OpenWRT router control plane is decoupled from the forwarding plane by creating a virtual bridge through the ovs-vsctl command. The location and port of the SDN controller are specified through virtual switch set-controller command and switch connectivity status changes to true. The interfaces that are targeted to control with SDN are associated virtually in ovs and a static IP address is given to each device connected in the network. The Backbone

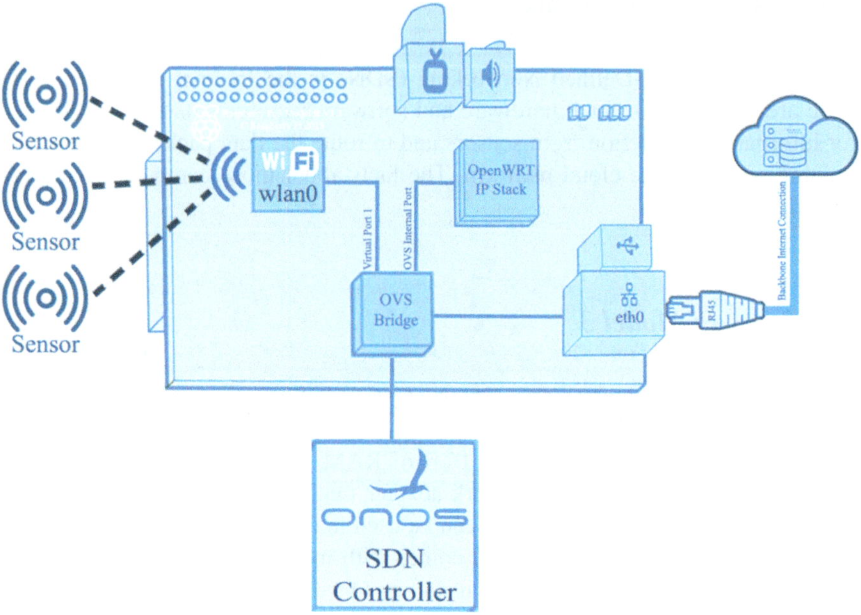

Fig. 13.1 Connection Setup of OpenWRT OVS in Raspberry Pi 3

internet connects to Raspberry Pi LAN port, and the IoT devices are associated with Wlan0 interface that was already set to AP mode and the devices start appearing on the ONOS controller GUI. Initially, when the devices are connected to the virtual bridge, they lose internet connectivity as the virtual network is not yet created and the connected devices are decoupled from the IP Stack. Once the reactive forwarding in the SDN controller is activated, the devices will start communicating with each other and after assigning the IP address, subnet, and default gateway the end devices can communicate over the internet.

ESP-32 LWS

ESP32 is reliable and trustworthy in industrial contextures, it is an open-source IoT Development board built on ESP-8266 wireless module with operating calefaction extending from −40 to +125 degrees centigrade. It can abolish external circuit defects and adjust to changes in external conditions embedded in Wearable, mobile electronic gadgets, and the Internet of Things applications, ESP32, achieve very low consuming power with a combination of various dedicated packages [7]. It also includes many features like overclocking, different modes, and wide ranges of power. ESP32 is immensely integrated with embedded network adapters. IoT applications using ESP32 require minimal circuit board designs.

LibWebSockets (LWS) is an easy-to-use, light-weight 1 MB package developed in C for executing advanced network protocols. By the integration of lws with HTTP/2, it generates a self-signed certificate when the device is booted for the first time supports secured data transmission over SSL and allows more intelligent management of packet streaming, compression of headers, priority and multiplexing queries [8]. The LWS allows to add up to four access points after flashing with esp-idf and connects automatically to available AP that is configured on boot up when other APs are down or unreachable. Once the devices are connected to the internet, it is possible to update the applications in the device Over the Air (OTA). The applications that run inside the device are different from the factory image and it is an isolated package that parts the internal disk into individual sectors for factory image and the applications. It is possible to implement the HTTP/2-based secure tunneling in LWS package [9], which adds extra layer security to the IoT devices. This also provides multiple benefits such as enhanced performance and memory consumption.

The esp-idf is a package that can be obtained from GitHub repository and flashed in ESP-32 device. The installation requires a 64-bit Linux toolchain, an integrated bundle that consists of all the necessary software tools. In case of windows, it is recommended to download and install MSYS2, which provides Linux environment in windows machine. It consists of a package manager (Pacman) to obtain all the required toolchain related to esp-idf. Before using the make in Linux it is required to install CMake version greater than 3.0 from the Pacman and genromfs package placed in the root directory and added to the system variables and upgrade the git repository. To flash the esp-idf, it is required to connect the device to the machine using USB and specify the /dev/ttyUSB0 or respectively connected USB port and execute the flashing command. Once the device boots up, it creates an access point named ESP to configure the device to deploy application and OTA update settings to monitor the serial port it is required to install termite RS232 terminal software.

This chapter mainly concentrates on priority-based packet stream management using HTTP/2 protocol in LWS framework. User code of LWS not be customized upon the using event loop by providing a generic timer of high resolution. It is accessible for Unix style platforms such as ESP32, Windows, Linux, and even BSD. LWS provides scalability to several thousands of connections at a time.

Open Network Operating System (ONOS)

Open Network Operating System (ONOS) was built on top of Java with Apache Karaf license and gives control panel to software network, network managing parts like links and switches, and software running programs to deliver communication services to adjoining networks and end hosts. It consists of management services for software-defined networks and customized communication routing [10]. Software that runs on this can be found in Use cases and Apps. Over multiple servers, it runs as a distributed system which allows using it in memory resources and CPU [11]. There are two types of membership for this, one is Partner and another one is Collaborator, with wavering stages of commitment. This software has been pre-owned as a platform, in which applications have been integrated into neighbouring projects or have been mentioned on top. It resolves all the disruptions in the software stack without failure by controlling the networking operation [12]. ONOS provides greater flexibility in handling data packets from the IoT devices and validating the data first by understanding the importance of the data. Also, the data path is controlled using SDN to route the data packets and manages the large-scale IoT devices in the SDN-enabled switch. In this project, an application was built on top of ONOS to enable priority-based critical traffic forwarding in IoT using LWS and HTTP/2 protocol in ESP-32 devices.

Sensors

IoT system has sensors or devices, which through some type of connectivity communicate with the cloud. Once the information reaches the cloud, it is processed by software and decides to function on actions like showing an alert or adjusting the sensors automatically without the user's need. Internet of Things also refers as the Internet of Everything, which consists of devices enabled with internet that gather, convey, and react on information which is given. Sensor data is the device's output that notices and answers to some kinds of input from the encompassing environment. Sensors are used to measure physical quantities. There are several types of sensors that are divided based on the quantities. They are Potential or radio or magnetic or Electric current sensors, Humidity sensor, Flow or fluid velocity sensors, Pressure sensors, Heat or Temperature or Thermal sensors, Proximity sensors, Optical sensors, and Position sensors.

As there are a wide-range of sensors, we can gauge all the physical changes in the external environment around us. In this project, we use a couple of elementary sensors that are broadly incorporated in each consistent day to day life such as DHT111, a temperature, and Humidity sensor, for producing real-time traffic in this testbed. It operates on power ranges from 3.5 to 5.5 V. It can determine 0–50 degrees centigrade of temperature with some precision of ±2 degrees centigrade and humidity which relatively ranges from 20–95% with the perfection of ±5%. It gives fully measured digital outputs for two measurements. It has its proprietary protocol of 1-wire and interaction between sensor and microcontroller, which is impossible with a direct interface of its peripherals. The protocol should be executed in the MCU firmware with accurate required timing by the sensor. This sensor consists of 4 I/O pins that are connected to Esp-32 wireless module General Purpose Input Output pins.

Related Works

In this chapter [13, 14], an architecture was predicted to support the small-scale proving ground to put SDN in the (IoT) Internet of Things. As it is pricier for acquiring the programming development boards like NetFPGA, the author builds up a testbed with Raspberry Pi and achieved Open vSwitch (OVS) with definitive protocol OpenFlow as a substitute productive setup. The products obtained are correlated with the outcome with both Raspberry Pi switch and NetFPGA-1G and proven that the network throughput is approximately the equivalent in both development boards. The advancement of data transmission technologies, smartphones, Nanoelectronics, cloud-based networking, things to the internet any place and at any moment, and physical devices enable connecting people. This is known as the (IoT) Internet of Things. At present, so many applications are being advanced for all round of IoT scenarios, like traffic control, industrial process monitoring, health care, home automation, civilian, earthquakes warning, and military operations. These applications in various domains claim their own committed systems and platforms. Most of the IoT services are supported by various systems and platforms, and a system or platform is correlated with specific application domains. This concludes in dedicated systems and platforms, individually consisting of many actuators and sensors, a group of gateways, a computing center, etc. In addition to this, IoT service providers reconfigure or design the protocols for communication and devices for every application and for every user. Also, they must manage a huge amount of inessential data. Software-defined network (SDN) technologies promote the building of network services. The reasonably integrated controller and the consistent southbound interfaces bring a key but dynamic architecture to provide basic functions and interfaces which can rapidly develop the user applications. Furthermore, incompatibility of data in various formats and models can be dynamically controlled over and easily achievable. Consequently, the applications in the various domains can be provided conveniently and accurately by the same system.

IoT structure consists of four components such as service, platform, network, and device; the service that supports the interface and communicates with the users, which connected with the platform to provide customized services to the users. There are several types of platforms in IoT which are data analysis, service development, device, and service platforms. For example, the device platform supports the execution of the services environment and development for users. The act of IoT platform is to support and execute IoT services. In the past, the platform was developed in a closed manner. However, industries and governments are refining assorted IoT platforms as open-source platforms. One of the reasons for this tendency is to improve the speed to get in the market, decrease the development cost and enhance the quality of software. By opening the software platform, many developers are given the opportunity to provide their effort. This will ultimately accelerate the development process and reduce cost. Moreover, the service development platform supports developing toolkits to users for them to develop IoT services easily. In the end, it provides the production and implementation of different applications. Including service and platform, one of the important components of IoT is the network. It helps to transmit data among devices, contents, services, and users. The network should have the capability to process, control and manage huge amounts of network traffic. In addition, security and privacy play a crucial role in IoT. Each stage of data transmission should have a security solution to secure it from errors. The basic components within the IoT service domains are apprehended behavior, the accomplishment of real-time awareness of the physical environment, and help with a human in charge through a wide range of determination and data visualization. IoT not only connects sensor devices and produce the data for an intention but also targets the computerization and expansion within the contemporary systems.

In addition, the data should be examined in a manner of acceptable modelling approach by understanding the significance of the data. This, the expected IoT services, such as health monitoring, smart homes, smart grids, smart cities, and smart traffic systems, have already integrated these fundamental models of computerization and resource expansion for any environment. The act of IoT devices is not defined to collect data, they should connect with assorted networks to provide a wide range of services. For example, IoT devices require the capability to interact either node-to-node or node-to-Web, which depends on whether the target node is in their own network. The flow for developing an IoT device is in open-source or open-hardware approaches for developers and manufacturers. For instance, IoT device production supplies basic standard architecture. With the basic standard or level architecture, users and developers can develop their own IoT devices.

In [15], an architecture with the horizontal model for developing real-time ready-to-use applications on the Internet of Things (IoT) using a software-defined network (SDN) technology was proposed. The author assesses the practicability to execute real-time IoT applications on top of the Cloud platform using Software-Defined Networking (SDN) in intermediate routers and gateways. The testbed was built up using Beacons (Bluetooth Radio Transmitters) that are associated with the Raspberry Pi switch operated by SDN controller in both north and southbound interfaces. The data exchange and the packet generation are made between smartphones that are connected. The information related to data exchange in SDN northbound and southbound interface through JSON format was explained clearly. In the JSON format packet data unit (PDU), priority is one of the elements which will be used as a base to route critical traffic in this chapter. An IoT (Internet of Things) architecture based on Software-defined Network (SDN) technologies facilitates allocating different resources, along with data, devices, and software, between different applications at various points. Incomparable data can be provided internally in the network, and advanced services and applications in various domains can be supplied accurately and rapidly.

OpenFlow is an approved interface which admits researchers to precisely control how packets are routes in actual Software-defined network (SDN). It is based on Ethernet switch but controls an open protocol which can be pre-owned to program the flow table in different routers and switches. And it also organizes three factors such as Flow table, OpenFlow protocol, and secure channel. Flow table contains the flow access that determines how data flows are handled in the network. Through flow entries, data flows can actively be adapted and broadcast in the network framework. A secure channel is pre-owned as communication means between the SDN switch and controller to organize a protective connection. OpenFlow protocol supports a basic interface that can be characterized externally by researchers, therefore eradicating additional programming in switches. The achievable ways to reduce the abeyance of critical traffic between two End-to-End IoT nodes in the identical network such as remote command robotic arm used during tele-surgeries as the traffic will be a crucial thing during the live operations. The simulation was accomplished using OpenDayLight SDN Controller, Mininet, and OpenFlow vSwitch. The results are appraised based on the ping and probe packets, but when a real-time critical data packet generates through the internet, many issues such as delay in MQTT Server response, congestion in border router due to other nodes, and Packet drop will arise due to connectivity errors.

In [16–18], it is explained that Wireless Sensor networks have a hidden motive of the Industrial IoT process and casting industrial automation-based applications. The development of IIoT accesses high potential for exfoliating device network and acquiescence with combined and regulated planning from low-level device activities to high-level significant communications of application data. This chapter delivers an industrial development desires program definability in the basic architectural components of IIoT, along with sensor cloud services, network infrastructures, wireless field devices, and IIoT gateways. Software-defined network is to deal with necessary challenges in an integrated IIoT system, such as accuracy, security, well-timed scalability, and quality of service. Advanced IIoT system design is expected based on the current networking technology like SDN scheduled to solve the critical traffic issues [16]. With the addition of IoT, communicability will not be barred to only smart devices. Based on Raspberry-Pi architecture, all the technical details about the IoT network in problems raised in industries are thoroughly analyzed.

The industrial growth in WSNs and installed systems has authorized several applications in various domains along with IoT systems. A distinctive class of Internet of Things authorized industrial management systems was proposed in IIoT, which maintains adequacy and economic advantages to install the system, maintainability, inseparability, and scalability. The complete design of IIoT systems and wide-ranging of IIoT applications desire an adjustable software definable efficiency for solving necessary challenges [17]. A new Software-defined IIoT design with its main segments depends on low power technical WSNs. Also, an expected architecture for software-defined activities in field devices, gateways, and sensor to send data to the cloud platform is not proposed. The quality of service can individually rule out the data flow programming equivalent to the active network conditions and adopt the SDN controller bond to asset the finest route for the actual time data communication. The network can be dynamically handled on critical data packet routing procedures and prioritization of service will be covered as part of this chapter.

The Problem Definition

In a very few years, the Internet of Things (IoT) has gone from innovation or set of advances that were cutting-edge to the circumstance today where incorporated in all the fields. If in case the IoT ecosystem has an issue or is presented to shortcomings, at that point the industries that are associated with it are similarly weakened. Truth be told while handling critical traffic is one of the serious issues affecting the improvement, there are various issues that stem specifically from this. IoT is transforming into its advancements as associated gadgets wind up more intelligent and progressively vivid, and desires to change over IoT information to a stream of knowledge to perceptions that boost economic value [19]. Also, the rapid adoption of IoT is driving the data acquisition and sensor costs down, allowing larger feasible business use cases that are been too costly earlier.

In IoT, connecting such a diverse number of devices is one of the greatest challenges and that will contradict the basic structure of current correspondence models lacks essential improvements. While the IoT yet possesses the ability to evolve into something in the prospects to improve the demand for businesses likewise, for those beforehand adopted this next generation technology still required to achieve somewhat in underlying network infrastructures. IoT had broadened its outskirts from the automated industrial sensors to wearable devices where the fundamental connecting network base is yet the same. The gadgets are intelligent enough to recognize changes in the encompassing environment however, the network devices are not quite intellectual to recognize the significance of the information they were routing. This makes a boundary between connecting network and innovation progression.

The Systems administration of extensive robotized machines is an ongoing concentration for modern computerization and one of the challenges is the availability with conventional network equipment that isn't intended to help more than connectivity of the local network. Modern systems can be exceptionally decentralized, inflexible and complex to oversee because of the tight coupling of the mechanization information and control plane that is regularly inserted inside the hardware. The registering and correspondence nodes are regularly arranged separately when the plant is set up and interconnections stay static from that point. The conventional mechanical interchanges progressive structure comprises of three system levels with different organizing advancements and conventions that limit what can be accomplished and includes multifaceted nature due to limited setup.

In the present IoT framework, we depend on centralized, server and client approach [20] using the SDN-enabled switches in which the servers are deployed on top of the Cloud platform and the client devices are IoT gadgets that are connected wirelessly and placed physically in local areas from where the data need to be gathered. The IoT client gadgets are associated with servers utilizing the internet. Figure 13.2 demonstrates a manufacturing industry containing different zones are furnished with multiple real-time sensors that are connected to the cloud server by the border gateway router to increase the overall productivity [21].

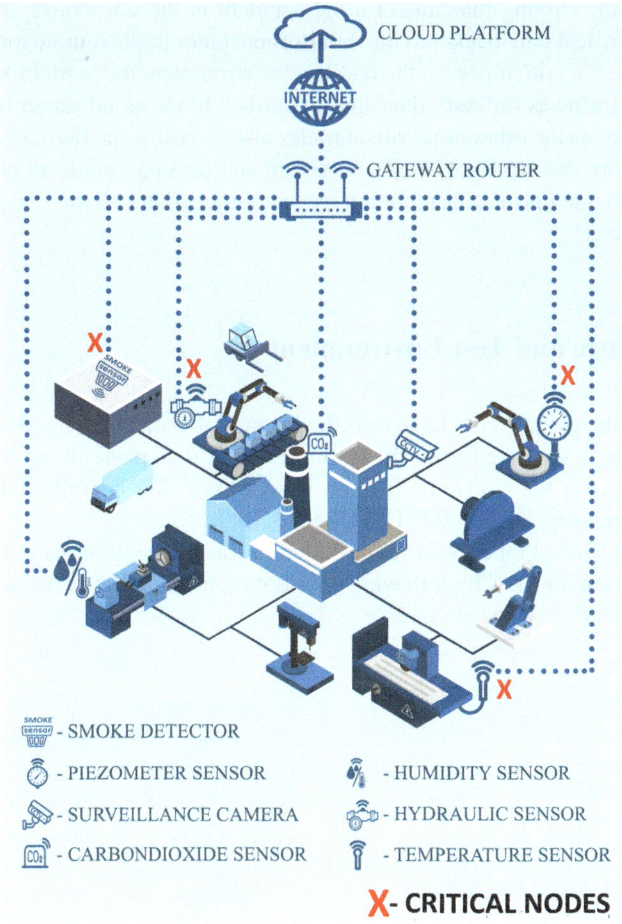

Fig. 13.2 Critical Nodes in an Industrial IoT Ecosystem

As we clearly observe the preceding smart industry there exist a few nodes in which the overall productivity is based on sensors that have a critical role and sensible towards significant data acquisition. Any suspension or delay of data movement from these sensors and the applications that are associated with it are thus impaired. The Hydraulic pressure sensor, temperature sensor fitted in carving machine, piezometer sensor in robot welding arm, and the smoke detector equipped in the warehouse are the crucial nodes that can arise a critical data packet at any moment. Alternatively, the surveillance camera, humidity, and carbon dioxide sensors stay broadly general-purpose nodes which are not that significant but connected to the same router.

At the point, if there is an emergency state in any one of the nodes in any zone, for instance, a decrease in the hydraulic pressure due to leakage or overheating problem in the carving machine or a fire accident in the warehouse, these sensors generate a critical data packet to the monitoring center in the remote location. Even though there is a criticalness in the real-time environment, the border router handles this critical traffic as ordinary data and forwards it to the cloud server this also creates the delay as the other non-critical nodes also overflow the border router simultaneously with their traffic [20]. These significant data packets from critical nodes need to be given a prior consideration by the border router so that the latency [19] can be reduced.

Architecture and Test Environment

To achieve the overall goal to reduce the latency of critical traffic, the OpenWRT control plane is separated from the IP Stack that is data plane in the border router [22] by utilizing the Software-Defined Networking (SDN) controller tool Open Network Operating System (ONOS) (Fig. 13.3).

OpenWRT-based Open vSwitch installed in Raspberry Pi remains as the southbound interface for SDN by following the steps to install in Linux that are displayed in OVS website.

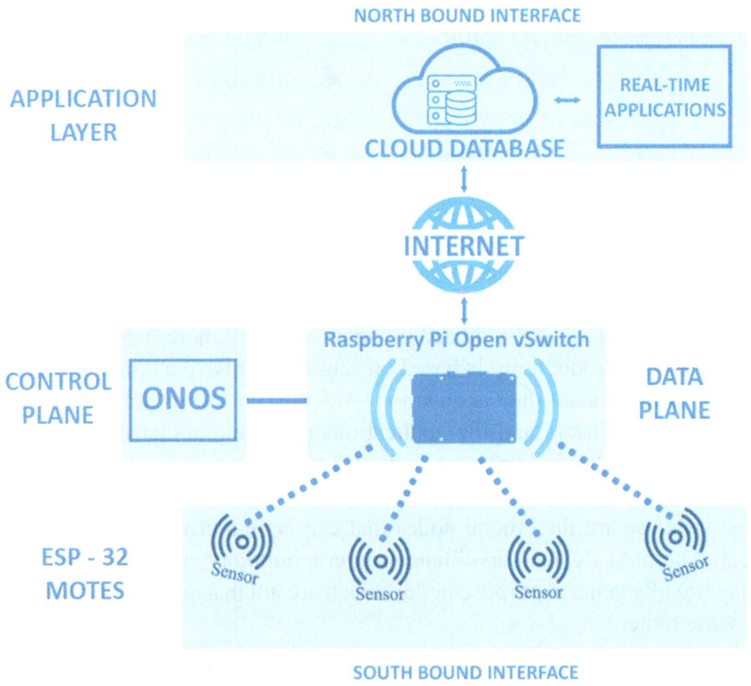

Fig. 13.3 Testbed Architecture

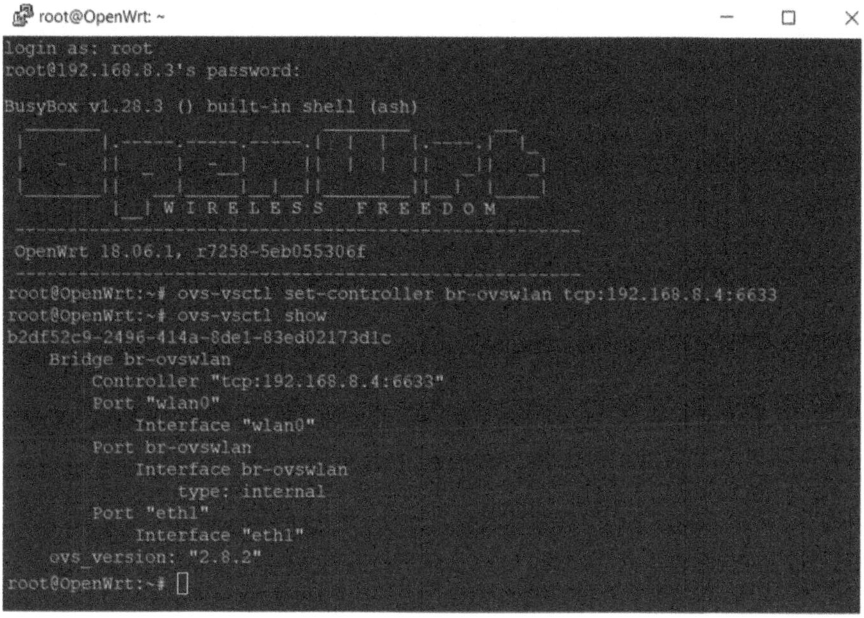

Fig. 13.4 OpenWRT Open vSwitch configuration

The Raspberry Pi's wlan0 was connected to a virtual port created in ovs-bridge named as br-ovswlan configurated in the Access point (AP) mode. The eth0 port is connected to backbone internet connectivity using RJ45 ethernet port in raspberry pi.

The virtual switch is interfaced with SDN controller using *ovs-vsctl set-controller br-ovswlantcp:ipaddress:port* as shown in Fig. 13.4 and once the controller is up and running the status can be seen through *ovs-vsctl show* command. The physical ports are added to the virtual bridge using virtual [22] port using the *ovs-vsctl add-port br-ovswlan wlan0*. When the ESP-32 devices are arranged in the wireless signal extent on boot up they are programmed to connect Pi access point automatically. If the esp devices are away from AP they search for other nearby APs that are configured through lws-based esp-idf in their range to transfer the data with the AP to the cloud server.

The ONOS is installed in a Linux PC or a VM operating on the Linux platform by following the steps mentioned in the ONOS website [6]. In case of controller installed on virtual machine, an additional virtual network adapter was attached in hypervisor and configured in bridged mode with external Wi-Fi adapter to allow OpenFlow standard protocol for packet delivery by enabling reactive forwarding application in onos controller and specify the flow rules so that the OVS running on OpenWRT Pi will start associating with onos and the topology of connected ESP devices in southbound devices will start to appear on the onos graphical user interface (GUI) (Fig. 13.5).

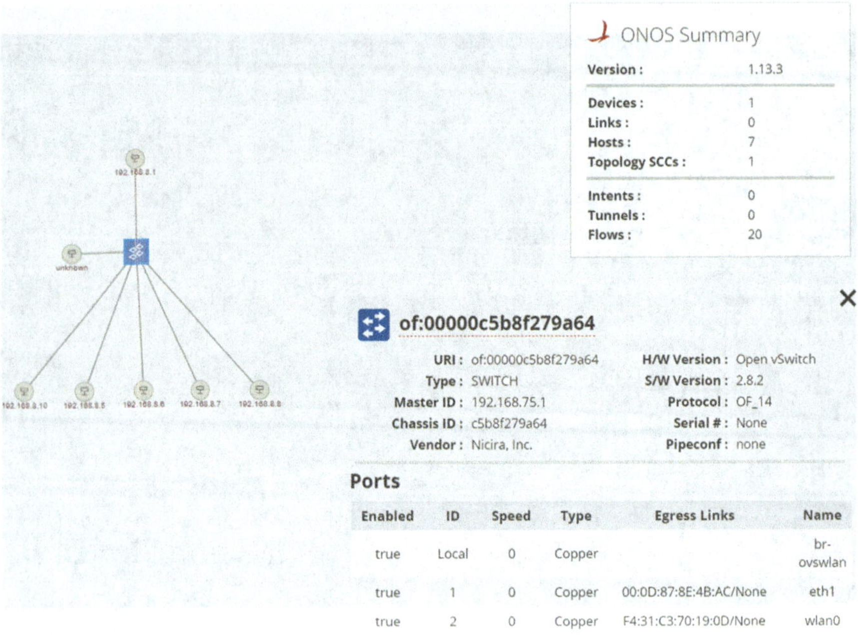

Fig. 13.5 Topology of IoT devices in SDN Controller

The onos controller requires the built-in ovs device driver application to be installed and started to interface with ovs-db. The GUI topology viewer displays the ovs switch and the number of devices connected to it. The devices are provided with a static IP address, as there is no concept of routing and switching in SDN, all the virtual devices are considered same to forward packets from the source to the destination [22] but onos also allows to use the remote DHCP server in the network using built-in DHCP Server app. The ovs switch is assigned an IP address and specify the default gateway to connect the southbound IoT devices to the internet. To allow the devices to start communication with the outside network the reactive forwarding application in onos is installed and started.

OpenFlow is the essential southbound protocol in the SDN control plane. OpenFlow works at time scales near the information arrange a round trip time (RTT), typically milliseconds, and is essentially used to program stream tables that control packet sending utilizing the match-activity display [23]. ONOS's sending observations principally use OpenFlow to program switch packet stream tables.

OpenFlow likewise gives some constrained configuration capacities, (for example, bringing a port up or down, or designing meters); however, many required setup activities [24], advising a change which controllers associate with, empowering the OpenFlow specialist on a switch, or empowering OpenFlow on explicit ports are unequivocally out of extension for OpenFlow.

Virtual switches that require arrangement past what OpenFlow gives may need to utilize other southbound conventions, for example, NetConf, ovsdb, SNMP, and so on [25]. ONOS right now gives drivers to a few southbound arrangement conventions. Later, ONOS may give setup reflections which enable gadgets to be designed in a non-exclusive, without agonizing over explicit conventions and their fluctuating semantics crosswise over gadgets. In this testbed we implemented OpenFlow v1.4 and the flow table obtained during the packet forwarding as shown in Figs. 13.6 and 13.7.

To monitor the real-time critical traffic from the IoT devices, a monitoring dashboard was created with some parameters which can measure the latency of critical traffic that has been generated from sensitive nodes in the IoT network. This dashboard also plots a real-time graph across time taken in milliseconds across time.

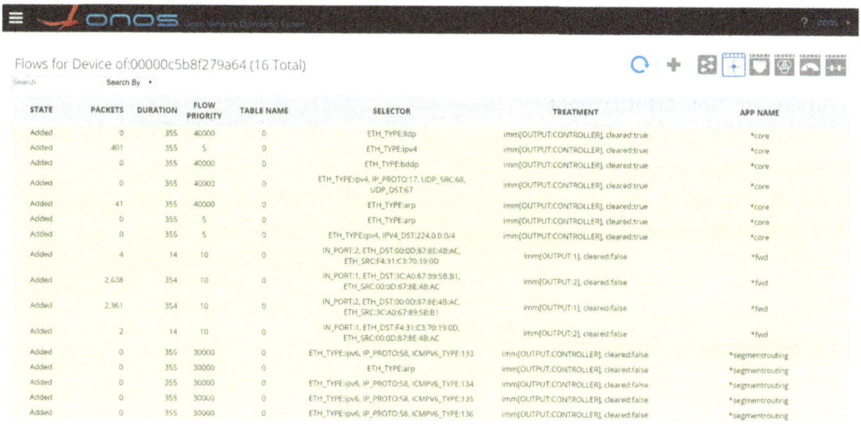
Flows for Device of:00000c5b8f279a64 (16 Total)

STATE	PACKETS	DURATION	FLOW PRIORITY	TABLE NAME	SELECTOR	TREATMENT	APP NAME
Added	0	355	40000	0	ETH_TYPE:lldp	imm[OUTPUT:CONTROLLER], cleared:true	*core
Added	401	355	5	0	ETH_TYPE:ipv4	imm[OUTPUT:CONTROLLER], cleared:true	*core
Added	0	355	40000	0	ETH_TYPE:bddp	imm[OUTPUT:CONTROLLER], cleared:true	*core
Added	0	355	40000	0	ETH_TYPE:ipv4, IP_PROTO:17, UDP_SRC:68, UDP_DST:67	imm[OUTPUT:CONTROLLER], cleared:true	*core
Added	41	355	40000	0	ETH_TYPE:arp	imm[OUTPUT:CONTROLLER], cleared:true	*core
Added	0	355	5	0	ETH_TYPE:arp	imm[OUTPUT:CONTROLLER], cleared:true	*core
Added	0	355	5	0	ETH_TYPE:ipv4, IPV4_DST:224.0.0.0/4	imm[OUTPUT:CONTROLLER], cleared:true	*core
Added	4	14	10	0	IN_PORT:2, ETH_DST:00:0D:87:8E:4B:AC, ETH_SRC:F4:31:C3:70:19:0D	imm[OUTPUT:1], cleared:false	*fwd
Added	2,638	354	10	0	IN_PORT:1, ETH_DST:3C:A0:67:89:58:B1, ETH_SRC:00:0D:87:8E:4B:AC	imm[OUTPUT:2], cleared:false	*fwd
Added	2,361	354	10	0	IN_PORT:2, ETH_DST:00:0D:87:8E:4B:AC, ETH_SRC:3C:A0:67:89:58:B1	imm[OUTPUT:1], cleared:false	*fwd
Added	2	14	10	0	IN_PORT:1, ETH_DST:F4:31:C3:70:19:0D, ETH_SRC:00:0D:87:8E:4B:AC	imm[OUTPUT:2], cleared:false	*fwd
Added	0	355	30000	0	ETH_TYPE:ipv6, IP_PROTO:58, ICMPV6_TYPE:133	imm[OUTPUT:CONTROLLER], cleared:false	*segmentrouting
Added	0	355	30000	0	ETH_TYPE:arp	imm[OUTPUT:CONTROLLER], cleared:false	*segmentrouting
Added	0	355	30000	0	ETH_TYPE:ipv6, IP_PROTO:58, ICMPV6_TYPE:134	imm[OUTPUT:CONTROLLER], cleared:false	*segmentrouting
Added	0	355	30000	0	ETH_TYPE:ipv6, IP_PROTO:58, ICMPV6_TYPE:135	imm[OUTPUT:CONTROLLER], cleared:false	*segmentrouting
Added	0	355	30000	0	ETH_TYPE:ipv6, IP_PROTO:58, ICMPV6_TYPE:136	imm[OUTPUT:CONTROLLER], cleared:false	*segmentrouting

Fig. 13.6 OpenFlow-based OVS Flows in SDN Controller

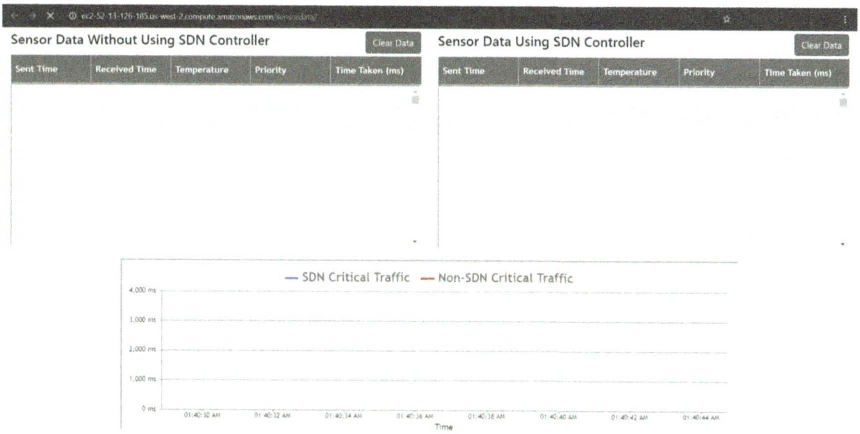

Fig. 13.7 Critical Traffic monitoring dashboard on Cloud Platform

The IoT devices are capable to obtain precisely synchronized time in milliseconds using Network Time Protocol (NTP) servers to make sure the accrued time at which the request was posted to a cloud database we have accessed the NTP Server services from ESP-32 devices with Unix EPOCH timestamp. The NTP services provide an option to specify offset time to match the local time zone. It was also observed that there is a slight delay about 1500 ms between the NTP and current time due to the UDP services in NTP causes a minor difference. To balance this variance the offset was increased to 1500 ms in IoT devices. The server is also capable of obtaining Unix time and the time taken is calculated from the difference between Sent Time and Received Time in milliseconds.

The real-time sample data was captured from IoT sensors and presented it on the Monitoring Dashboard. The parameters considered are Sent Time, Received Time, Sensor value, and Priority. Time Taken is calculated based on these parameters. As shown in Fig. 13.8, the parameters are obtained from the Non-SDN nodes and it was observed that the Time Taken in milliseconds for High priority data is the same compared to non-priority data.

In case of data from SDN-incorporated devices, the Time Taken in milliseconds is reduced even on same internet speed.

As shown in Fig. 13.9, the Time Taken for high priority data is less compared to Non-SDN device critical traffic.

The Proposed Solution

The main intention is to reduce the abeyance of critical traffic. First, we will analyze the central nodes in the network that can trigger critical state at any time (Fig. 13.10).

Sensor Data Without Using SDN Controller Clear Data

Sent Time	Received Time	Temperature	Priority	Time Taken (ms)
01:32:58	01:34:01	39.44	LOW	3316
01:33:00	01:34:04	39.44	LOW	3329
01:33:02	01:34:06	52.78	HIGH	3303
01:33:05	01:34:08	39.44	LOW	3429
01:33:07	01:34:11	38.89	LOW	3160
01:33:10	01:34:13	39.44	LOW	3223
01:33:12	01:34:16	52.89	HIGH	3125
01:33:15	01:34:18	39.44	LOW	3301

Fig. 13.8 Sample Data of Non-SDN Traffic on Monitoring Dashboard

Sensor Data Using SDN Controller

Clear Data

Sent Time	Received Time	Temperature	Priority	Time Taken (ms)
01:33:00	01:34:04	50.78	HIGH	2094
01:33:02	01:34:06	37.44	LOW	2635
01:33:05	01:34:08	50.44	HIGH	2369
01:33:07	01:34:11	35.89	LOW	2542
01:33:10	01:34:13	37.44	LOW	2658
01:33:12	01:34:16	49.78	LOW	2645
01:33:15	01:34:18	50.78	HIGH	2122
01:33:17	01:34:21	36.44	LOW	2541

Fig. 13.9 Sample SDN Data Captured on Monitoring Dashboard

Every IoT node in the network will have inferior and superior sensor values. For example, the inferior for heart rate monitoring sensor is 55 beats per minute and superior is 105 beats per minute. When the sensor reports a value that is external of this range, then that will be deliberated as a critical state [26]. At first, the priority value will be set to zero and when there is a critical state the priority did replace one in the packet header. The sensors are graded to the limits to set priority depending on the obtained values and move onward to the SDN controller.

```
functiondispatchPacketData ()
  priority = 0
  inputsensordata
  loop while sensordata # 0
   ifsensordata>= 105 OR sensordata<= 55 then
     priority = 1
   end If
   if priority > 0 then
     path = MinimumTimeDelay
   else
     path = Regular
   end If
   outputCloudDatabase
  end while
end function
```

The controller oversees the packet priority and determines the path in which the data packet requires to be routed. When the priority of a data packet is exceeding zero, then the SDN controller will appreciate the imperativeness of that packet

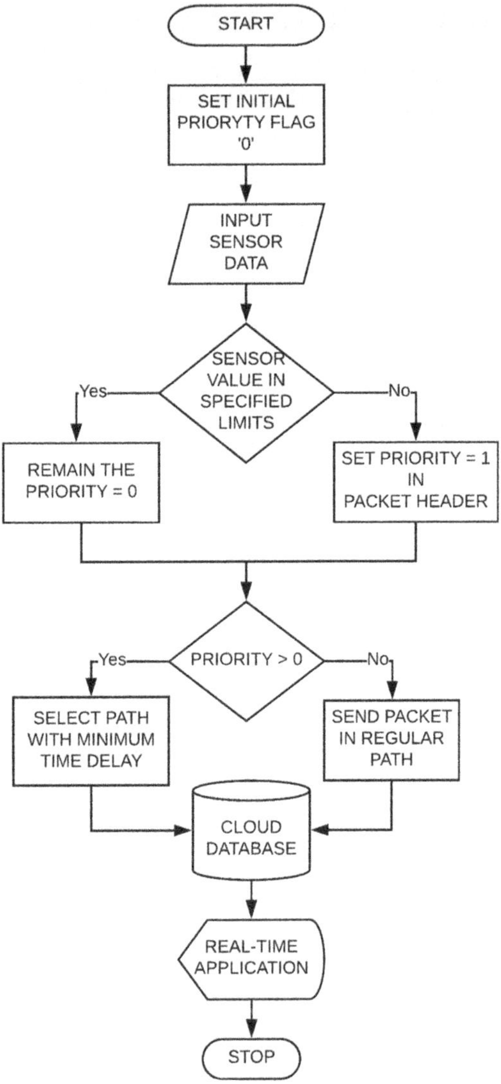

Fig. 13.10 Flowchart to Route Critical Traffic

requires to be paid distinguishing attention and routes in a path with the least deferment to the cloud database where the users will get real-time notifications in applications [27]. In case of priority zero, the SDN controller will be ahead of the data packet in the usual path. In this manner, by outlining the critical nodes in the IoT network, latency of the critical traffic can be declined.

In HTTP/2 the data is transmitted in streams and bit 5 in the header is an unsigned 8-bit entity staging a PRIORITY frame that can be posted on a stream in multiple states; however, it is not possible to send within continuous frames that include one header frame [9]. There is a possibility of reaching those frames post-processing of data, which would make it have no impact on the distinguished stream which doesn't influence outline transmission on that stream. It is functional for a stream to end up shut while prioritization data that makes an impact on that stream is in transmission. Since the stream could be given a need that is not the same as what is planned. To maintain a strategic distance from these issues, an endpoint should hold stream prioritization state for a period after streams become shut. The more drawn out state is held, the lower the opportunity that streams are selected wrong priority or default priority of the streams.

The PRIORITY of frame can be sent for a stream in the shut or idle state. This considers the re-prioritization of a grouping of child streams by adjusting the need of an unused or shut parent stream. The payload of a frame contains a header block fragment. Prioritization of data in a HEADERS outline is sensibly equal to a different PRIORITY outline; however, incorporation in HEADERS avoids the potential for agitating in stream prioritization when fresh streams are made. Prioritization fields in HEADERS outlines resulting to the first on a stream re-prioritize the stream.

Result and Discussion

By implementing the suggested testbed architecture with the proposed solution, the latency of traffic in the critical IoT nodes can be minimized. To observe these changes, a real-time critical state can be created in any of the IoT sensors connected to the southbound interface of the SDN Controller, and the packet propagation time can be traced using the packet monitoring tools like Wireshark. Also the Unix EPOCH of the packet sent is differentiated with time reached in the Cloud server network interface, and the average time taken to propagate critical traffic to Cloud database can be observed in the created monitoring and compared with the time taken without using SDN controller by means of different network speeds.

As presented in the Fig. 13.11, the graph compares the latency of critical traffic in IoT with SDN and without using SDN. The latency using SDN traffic is observed to reduce with respect to time contrasted to normal traffic. This result was posted on an average speed of 30 mbps. It was also tested across various internet speeds on an average scale in which the latency is constantly less compared to Non-SDN Critical Traffic. It was also observed that packet stream using HTTP/2 priority states does not affect the regular traffic produced by the other nodes in the network.

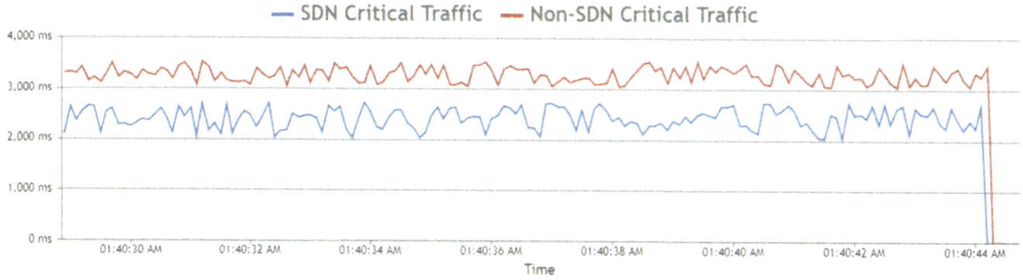

Fig. 13.11 Flowchart to Route Critical Traffic

Conclusion

In the current system, an eternally rising number of difficulties assorted real-time applications in IoT that are operating on top of the cloud platform are addressed. SDN innovation is employed to accomplish an excellent concept among the possible number of open-source techniques to send critical data from one point to another in the system. This chapter discussed the adversities that arise in IoT while managing emergency traffic during the critical state. By having those arguments in mind, the SDN technology is included in the system assuming that it is one of the decisive among the existing solutions. When the priority label in the HTTP/2 packet header can be changed with an account to time and the sensor value, it is easy for an SDN controller to route the packet with least latency. The procedure that is expected in the system to determine the critical nodes in the IoT network is taken as an advance by implementing practically on a testbed as the following work. The expected way does not supply any excess additional security mechanism to the IoT system in SDN which can be introduced as a forthcoming assignment [28, 29]. Our upcoming intention is to implement major algorithms and functions and the security-related issues [30] in the SDN controller on an IoT network.

Acknowledgments I would like to express my wholehearted thanks to my educator (Mr. K. Venkatesh) who gave me this elegant chance to explore new trending innovations and to the organization of this research work that helped me gain knowledge in cutting-edge technologies.

Besides, I would like to thank my companions for giving intense support to conclude this task within a limited period.

References

1. Gartner, Inc., Gartner Says 13.6 Billion Connected "Things" Will Be in Use in 2018, Up 30 Percent From 2015. Press Release, Stamford, Conn., November 10, 2015.
2. Dores, C., Reis, L., & Lopes, N. (2014) Internet of things and cloud computing. In: *2014 9th Iberian Conference on Information Systems and Technologies (CISTI)*.
3. Venkatesh, K., & Nagarajan, V. (2018). A framework to support dynamic routing and traffic provisioning in optical networks using virtualization and software-defined networking. *International Journal of Pure and Applied Mathematics, 118*(11), 733–737, ISSN: 1311–8080.
4. Bera, S., Misra, S., & Vasilakos, A. (2017). Software-defined networking for internet of things: A survey. *IEEE Internet of Things Journal, 4*(6), 1994–2008.
5. Setup – Raspberry Pi Documentation. Raspberrypi.org, 2018. [Online]. Available: https://www.raspberrypi.org/documentation/setup/. Accessed 09 Sept 2018.
6. Open vSwitch Documentation – Open vSwitch 2.10.90 documentation. Docs.openvswitch.org, 2018. [Online]. Available: http://docs.openvswitch.org/en/latest/. Accessed 09 Sept 2018.
7. Get Started – ESP-IDF Programming Guide v4.0-dev-141-g106dc0590 documentation. Docs.espressif.com, 2019. [Online]. Available: https://docs.espressif.com/projects/esp-idf/en/latest/get-started/. Accessed: 25 Mar 2019.
8. warmcat/lws-esp32-factory. GitHub, 2019. [Online]. Available: https://github.com/warmcat/lws-esp32-factory. Accessed 25 Mar 2019.

9. Hypertext Transfer Protocol Version 2 (HTTP/2). Http2.github.io, 2019. [Online]. Available: https://http2.github.io/http2-spec/. Accessed 25 Mar 2019.
10. ONOS – Home – Wiki. Wiki.onosproject.org, 2018. [Online]. Available: https://wiki.onosproject.org/. Accessed 09 Sept 2018.
11. Kandah, F., & Schmitt, S. (2018). SAND: Smart and adaptable networking design using virtual slicing over software-defined network. *EAI Endorsed Transactions on Internet of Things, 4*(13), 155333. https://doi.org/10.4108/eai.6-4-2018.155333.
12. Anand, N., Babu, S., & Manoj, B. (2018). On detecting compromised controller in software defined networks. *Computer Networks, 137*, 107–118. https://doi.org/10.1016/j.comnet.2018.03.021.
13. Kim, H., Kim, J., & Ko, Y. (2014). Developing a cost-effective OpenFlow testbed for small-scale Software Defined Networking. In: *16th International Conference on Advanced Communication Technology*.
14. Hu, P., Ning, H., Chen, L., & Daneshmand, M. (2019). An open internet of things system architecture based on software-defined device. *IEEE Internet of Things Journal, 6*(2), 2583–2592. https://doi.org/10.1109/jiot.2018.2872028.
15. Li, Y., Su, X., Riekki, J., Kanter, T., & Rahmani, R. (2016). A SDN-based architecture for horizontal internet of things services. In: *2016 IEEE International Conference on Communications (ICC)*.
16. Ahmed, K., Blech, J., Gregory, M., & Schmidt, H. (2018). Software defined networks in industrial automation. *Journal of Sensor and Actuator Networks, 7*(3), 33. https://doi.org/10.3390/jsan7030033.
17. Muciaccia, T., & Passaro, V. (2017). Future scenarios for software-defined metro and access networks and software-defined photonics. *Photonics, 4*(4), 1. https://doi.org/10.3390/photonics4010001.
18. Li, J., Yang, L., Fu, X., Chao, F., & Qu, Y. (2017). Dynamic QoS solution for enterprise networks using TSK fuzzy interpolation. In: *2017 IEEE International Conference on Fuzzy Systems (FUZZ-IEEE)*.
19. Meseguer Llopis, J., Pieczerak, J., & Janaszka, T. (2016). Minimizing latency of critical traffic through SDN. In: *2016 IEEE International Conference on Networking, Architecture, and Storage (NAS)*.
20. Banerjee, A., Sufyanf, F., Nayel, M., & Sagar, S. (2018). Centralized framework for controlling heterogeneous appliances in a smart home environment. In: *2018 International Conference on Information and Computer Technologies (ICICT)*.
21. Srinivas, L. N. B., & Ramasamy, S. (2017). An analysis of outlier detection techniques for wireless sensor network applications. *International Journal of Pure and Applied Mathematics, 117*(16), 561–564, ISSN: 1311–8080..
22. Zhou, W., Li, L., Luo, M., & Chou, W. (2014). REST API design patterns for SDN northbound API. In: *2014 28th International Conference on Advanced Information Networking and Applications Workshops*.
23. Sayee Kumar, M., & Purusothaman, T. (2015). Hybrid broadcast group management protocol for secure, scalable and efficient group communication. *Journal of Computer Science, 11*(2), 344–350, ISSN: 1549–3636..
24. A framework of controller with flow table cache and performance analysis in software defined industrial networks. RevistaTecnica De La Facultad De Ingenieria Universidad Del Zulia (2016). https://doi.org/10.21311/001.39.3.27.
25. Yu, H., Ng, B., & Seah, W. (2018). TTL-based efficient forwarding for nanonetworks with multiple coordinated IoT gateways. *IEEE Internet of Things Journal, 5*(3), 1807–1815. https://doi.org/10.1109/jiot.2018.2812868.
26. Venkatesh, K., Srinivas, L., Mukesh Krishnan, M., & Shanthini, A. (2019). QoS improvisation of delay sensitive communication using SDN based multipath routing for medical applications. *Future Generation Computer Systems, 93*, 256–265. https://doi.org/10.1016/j.future.2018.10.032.

27. Bisio, I., Lavagetto, F., & Luzzati, G. (2016). Cooperative application layer joint video coding in the internet of remote things. *IEEE Internet of Things Journal, 3*(6), 1418–1426. https://doi.org/10.1109/jiot.2016.2610103.
28. Karthik, G. M., Sayee Kumar, M., & Aruna, A. (2017). Fully Homomorphic public key encryption based on arbitrary key aggregation cryptosystem for secured data communication in cloud infrastructure. *International Journal of Pure and Applied Mathematics, 117*(15), 1143–1149.
29. Flauzac, O., Gonzalez, C., & Nolot, F. (2015). Original secure architecture for IoT based on SDN. In: *2015 International Conference on Protocol Engineering (ICPE) and International Conference on New Technologies of Distributed Systems (NTDS).*
30. Vishwesh, J., & Rajashekar, M. B. (2017). Internet of Things (IoT): Security analysis & security protocol CoAP. *International Journal of Recent Trends in Engineering and Research, 3*(3), 417–425. https://doi.org/10.23883/ijrter.2017.3126.6ibua.

Chapter 14
Integrating IoT and Machine Learning – The Driving Force of Industry 4.0

A. Suresh, R. Udendhran, and M. Balamurugan

Explosive Growth of Intelligent Manufacturing in Industry 4.0

Intelligent factories are created by the initiative action taken by the German strategic Industry 4.0 in which the cyber-physical systems (CPSs) play a role in updating and transforming the manufacturing technologies by cloud computing and Internet of Things (IoT). In the era of Industry 4.0, the physical process is being monitored by the manufacturing system called as "digital twin" in the physical world, which communicates with the machines, sensors and humans and make smart decisions by communicating in real time. Industry 4.0 is a combination of intelligent production processes and the embedded production system, which could make transformation from the production and industry value chains as well as the business frameworks.

In Industry 4.0, updating the manufacturing systems to an intelligent level takes place and that makes possible the availability of manufacturing technologies and sophisticated information in order to achieve a smart, flexible and reconfigurable process in manufacturing to look over the global and the dynamic market as shown in Fig. 14.1. Large companies and industries, small- and medium-sized enterprises (SMEs), and supply chains for holistic manufacturing require certain information and physical process flows that are enabled by Industry 4.0. Based on the varying

A. Suresh (✉)
Department of Computer Science and Engineering, Nehru Institute of Engineering and Technology, Coimbatore, India

R. Udendhran · M. Balamurugan
Department of Computer Science and Engineering, Bharathidasan University, Trichy, India
e-mail: mbala@csbdu.in

© Springer Nature Switzerland AG 2020

G. R. Kanagachidambaresan et al. (eds.), *Internet of Things for Industry 4.0*,
EAI/Springer Innovations in Communication and Computing,
https://doi.org/10.1007/978-3-030-32530-5_14

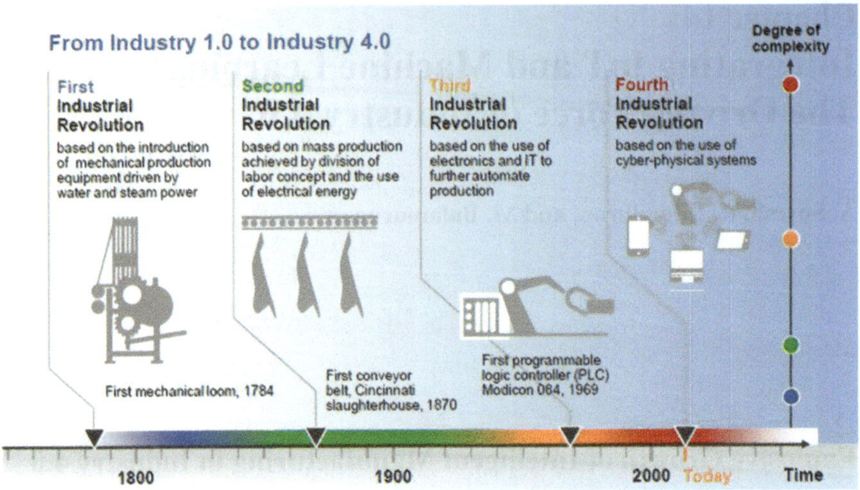

Fig. 14.1 Development of intelligent industry as presented in AI business

situations and necessities, certain keystone technologies are required for the intelligent manufacturing process, based on the previous behaviour of the changing nature. Therefore, adaptive decisions could be made by solving the problem through direct communication among the manufacturing systems in a timely manner. In certain cases, Artificial Intelligence (AI) could be adapted by certain technologies that learn from the past experience with the intelligence, co-operated and ever-present practice in industries [1].

Machine Learning Is Expected to Drive Growth in Industry

Intelligent manufacturing also includes IoT-enabled services with the cloud computing technology for storage purposes. Nowadays, the amount of data generated and its use are increasing rapidly day by day [2]. The data generated contains various format, qualities and semantics, for instance, the data collected from the environment, production and machine lines. This is mentioned in different forms, namely Industry 4.0 at Germany, Smart factory in South Korea and Smart Manufacturing in USA. The quality-related data possibility has the potential to enhance the quality of the product and process sustainability [3]. The surface roughness could be determined with the help of the statistical features of the machine learning approach. However, machine learning possess certain negative impacts as well as challenges, since in certain cases this could distract the system from the main issues or challenges and lead to conclusions that in turn lead to incorrect actions [4, 5].

Fig. 14.2 Machine learning in automobile industry

For the transformation in manufacturing domain could be made by the collaboration of developments made in mathematics and computer science (e.g. statistical learning) with the easy–to-use tools offered freely to grasp and sustain the repositories of increasing manufacturing data in automobile industry as shown in Fig. 14.2.

Machine Learning in Industry 4.0

Predictions on machine learning technologies can be made based on the wide range of data available [6]. This could draw knowledge from the experience independently and construct pattern recognition. But in terms of the future, AI has no scope in the long term [7]. Internet of things are believed to be distant in concepts due to their enormous space for storage and the branches of AI, machine learning and deep learning makes use of big data for the optimization process, gain novel solutions and obtain novel insights. Companies and industries ranging from small to large international sectors all require data that could be used by them. Predictions could be made with the help of the available software that deals with evaluation and consolidation [8]. Process efficiency could be obtained with the new solution by properly analyzing the data obtained from the sensor, customers and logs [9]. Moreover, an IT infrastructure is required due to the addition of huge amount of data, which is attuned to machine learning and AI workloads and process as shown in Fig. 14.3. The main tasks of the machine learning are as follows: (1) recognizing patterns and (2) arriving at solutions from them. Further, these findings could be utilized at their best [10].

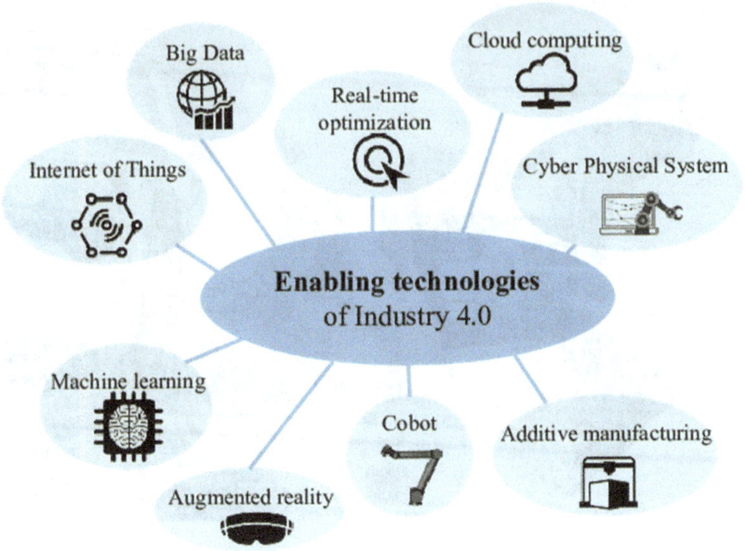

Fig. 14.3 Machine learning as enabling technology in Industry 4.0

Current Applications of Machine Learning in Smart Industries

Image recognition is the commonly used learning method at present. Certain other processes include intelligent bots or digital assistants, speech processing, speech recognition, face recognition, analysis of text and video, and automatic translation and transcription.

Artificial Intelligence in Industries 4.0

In smart factories, all the components are interconnected, i.e. the machines, components and the interfaces could communicate with one another. The manufacturing process could be optimized with the huge amount of data. With the help of image recognition and analysis, optimization of big data could be made, which serves as the best example.

Digital Monitoring and Control

In production industries, the system can identify the objects on the conveyor belt and can sort them all as shown in Fig. 14.4. In quality control, this intelligent system works well detecting flaws and colour changes in the product.

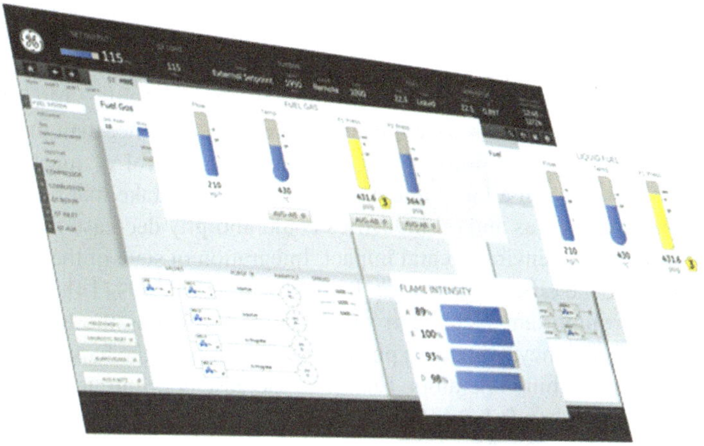

Fig. 14.4 Monitoring with high-performance HMI/SCADA

Artificial Intelligence and Predictive Maintenance

Machine learning techniques are highly used in the field of support services and maintenance of companies [11]. The maintenance of cycle, energy consumed from the individual machines could be captured by AI and these information deals with optimization of the process in the concurrent stages. The need for replacement of the parts and the defective areas could be indicated by the operating data. The increasing amount of data makes the optimization process good with more accurate predictions.

Brief Application of Deep Learning in Industry

Deep learning (DL) gains the attention over commercial machine learning process by two factors in the field of prediction as well as at the training [12]. (1) For the process of training, the features needed for engineering and handcrafts requirements are highly reduced [13]. (2) Extraction could be done the easiest way by DL techniques, which are not caught by the human view. Moreover, accuracy is also enhanced by the DL techniques. Based on the statement given by experts, the deep learning process tends to increase the growth of the economy. The maximum benefits are accepted in the field of IT, telecommunications, finance sectors and in manufacturing companies.

Applications of Human Machine Interface (HMI) in Industry 4.0

Applications of HMI gain attention in consumer use cases like the virtual assistants as shown in Fig. 14.5. Necessary applications are focused rather than the repetition of messages that tends to impact the industries. The control of machines in the industries is said to gain a lot of attention. Plague or a mistake that happens at the user interfaces in the gas and oil industries could abruptly decrease the production rate and may have an environmental impact. Integration of state of the art quality in HMI could raise the efficiency level at the field with new levels [14].

Safety is an important factor that should be concerned as the highest priority rather than the efficiency. Obtaining physical feedback from the haptic technologies is beneficial for machine operators. Enable to increase concentrate on the primary task by freeing the hands by AR or wearable glasses. Real-time status from the machine could be obtained by the operators so that the prioritization of workloads with the requirement of material and anticipating tool could be gathered at every individual task. Off-site monitoring could also be done by the specialized agents that could enhance the safety around the environment or else it could require the

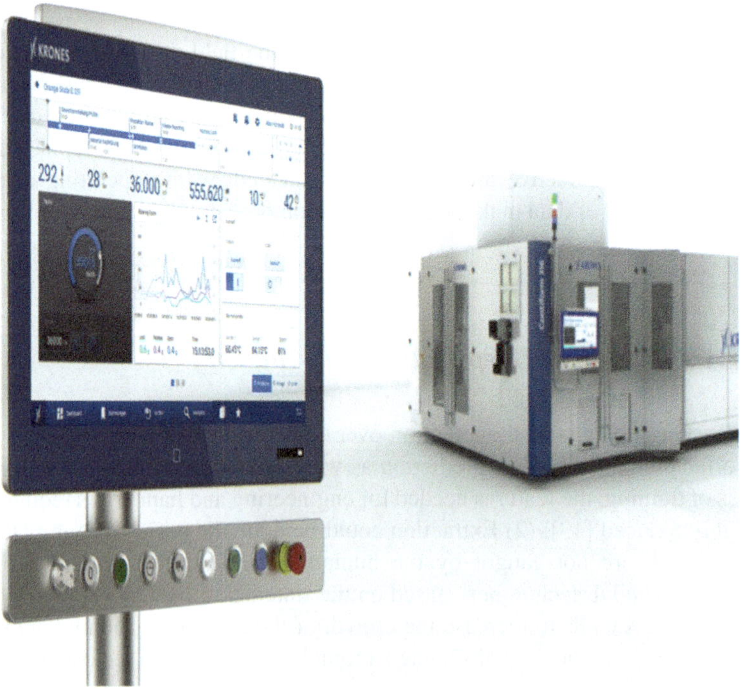

Fig. 14.5 Combined HMI in industry

labor or concerned person to walk to that particular locality. Surveying the status could be done by the promoted managers of that specific plant, and when a bottleneck around the field occurs, the immediate notifications could be received from the user end [15].

New HMI technology also includes automotive sector as a rapid adopter. In the high-end model of the infotainment systems, new HMI technologies are determined, and the application also includes the redesign of mundane features like door openers. The technologies most desired by the automakers include the gesture recognition and touch screen technologies. Touch screen in the car will not be same as the one present in the smart phones. This is due to the fact that capacitive screens are being replaced by the resistive screens in most of the mobile phones. This is done to lower the cost, increase the longevity etc., enable to increase the adaption of new technology. Advanced Driver-Assistance Systems include infrared sensors that could help in monitoring the driver's attentiveness [16].

The car includes gesture recognition that is fitted at the outside. When drivers who carry held load of groceries in their hands approach their Ford Escape car parked, they could open the tailgate by passing their foot underneath the rear bumper. In Digital Out of Home ("DOOH") advertising, advertisers tend to use the concept of gesture recognition and eye-tracking technology. Deployment of digital displays is highly expensive. For sale purposes, ad displays could be fitted in the form of wall-mounted advertisement. In certain other cases, through downloading the application other information could be obtained. The combinations of new analytics with the higher engagement that enable advertisers to aim specific group of users have resulted in new revenue for both media companies and their customers alike.

In recent advancements, the connected devices ("Things") with their virtual representation have gained more prominence than ever, creating effective applications, services like smart healthcare, intelligent industry 4.0, smart transportation for less pollution with environmental monitoring. These areas have been of focus for many researchers and organizations as well as standardization bodies among the communities in semantic web and insurance companies. The main success behind IoT is that it provides flexibility to applications that demand communication, storage, accessibility and service platforms and create opportunities for important sectors such as retail, green energy, manufacturing, smart homes as well as personalized end-user applications. Therefore, it is evident that IoT plays an important role in daily routine. Generally, the huge volume of data generated from industries, internet is overwhelming and reaches 2.5 quintillion bytes of data and in past several years, around 90% of data was generated from various sources. Among these data, the data generated from the sensors known as "Sensory data" provide meaningful information when analyzed by employing algorithms and they can be converted into a better understanding about real human world leading the machines to act smartly and learning by themselves which is known as "Artificial Intelligence". Artificial Intelligence offers automation to application, products and knowledge hierarchy and provides effective data transformation in which we can adapt meanings to the stages based on the context of semantics and IoT as shown in Fig. 14.6.

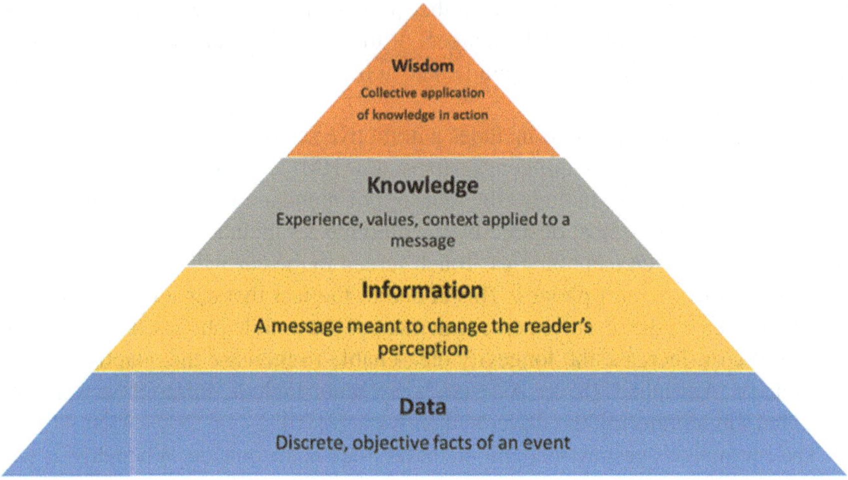

Fig. 14.6 Knowledge hierarchy in the context of IoT

Machine-Learning-Based Human Machine Interface in the Context of Industry 4.0

Machine-learning-based HMIs are the recent trend that has been going on in all the process industry 4.0. This HMI could have an access through the web, which means that any device could be connected once they gain their access through the internet. These HMIs are a great solution to the traditional HMI, and we do not need any human intervention [17]. User would be able to monitor and control the devices from a remote location. It is not necessary for the user to be physically present. With the rising use of mobile application, this could be made easier with the go on process and get the notifications by hand [18].

Since these HMIs rely on internet access, updates could be done automatically and ensure the system is using the latest notifications. As soon as the user logs in to the application, it could send the most recent updates for the users, relevant of accessing the usual webpage [19].

One of the major problems that occur in these HMIs is security issues. Privacy is the greatest threat for internet-based applications. A proper connection should be maintained among the components of the HMI connection in order to enhance the reliability. A hacker could attempt to gain access over the component and change the configuration, which could interrupt all the other systems connected to it [20]. This is extensively done in DoS, since hacking the server could interrupt all other nodes connected to it. During the protocol development, it is necessary for the protocol designers to pay attention to this threat.

Multiple industries are involved in this machine-learning-based human machine Interface for the effective operation of the HMI systems as shown in Fig. 14.7. An issue known as federated identity management occurs, which needs the access from

Fig. 14.7 Example of machine-learning-based human machine interface in Industry 4.0

the remote organization for the multiple components involved in the HMI since several equipment need certain authorization (person or equipment involves in getting permission from other entities). This issue could be solved by Security Assertion Markup Language (SAML), Web Services Trust (WS-Trust), and PKI [21]. It is necessary to have interoperability, since no conflicts should happen among the involved industries.

This web-based HMI is not required in every field. Local HMIs could pay a severe attention where there is a necessity of a person to be physically present in that environment. Each organization should finely balance the needs [22]. If we opt to gain mobility, convenience and monitoring, then we should use only the web-based application. Otherwise the traditional HMIs are the best.

In-Depth Analysis of Machine Learning

The Internet consists of thousands of machines, instruments, equipment, drones, consumer electronics, wares, utilities, robots and then millions of sentients, intelligent and digitized objects. That is, the current Internet, which simply consists of computer servers, desktops, laptops, smartphones and hand-held, is bound to scale exponentially to have all kinds of devices and things. Precisely speaking, the current Internet is the network of networked computers [23]. But the future is the network of networked computers, devices and digitized entities. In other words, the Internet of devices, services, energy and things is to emerge and evolve rapidly to empower not only businesses but also people in their daily deals, decisions and deeds. When such kinds of connected and integrated devices, services, data sources, applications, and things communicate with one another, there will be massive amounts of interactional, transactional, commercial, operational and analytical data. Therefore, IT professors and professionals across the world are striving hard and stretching further to bring forth pioneering data analytic products, platforms, processes, patterns to meticulously capture, cleanse and crunch the growing data volumes using the batch as well as real-time processing methods to extricate actionable insights in time. Thus, IoT data along with its analytic can result in smarter and sophisticated applications not only for worldwide businesses but also every individual in the increasingly connected countries, counties and cities [24]. In this technological world computers have made their footprints in every field. Without the computer, technology does not exist, and hence with the evolution of computers we are able to communicate at a faster rate, for instance e-commerce and on-line marketing have found their success in many developing countries in which internet is the backbone in powering these technologies, in the case of industries, which needs computer to operate their machines as well as store and retrieve data. A lot of equipment are calibrated with the help of the computers. Nowadays storage of large amount of data is possible with the help of big data technologies [25]. In the present scenario, machine learning seems to be trending, which makes the machine to learn by itself and integrating it with the Internet of Things. For instance, IoT-based healthcare provides direct connection to equipment (MRIs) and uses Artificial Intelligence for the initial triage to diagnose problems and reduce the requirement of putting someone on site by 40%, thereby reducing response time and improving customer satisfaction. Healthcare companies are streaming EMR data and using deep learning to predict the health of patients that are potential for code blue or sepsis shock and send a nurse to visit. Technology has been increasing day by day. Machines play a vital role in reducing the effort of humans. It is necessary to have a connection to all those machines. Machine learning algorithms have three distinct phases: learning, validation and classification [26]. The first phase is when the algorithm learns to classify data points by creating a knowledge model with the most relevant data features. Machine learning algorithms take annotated data as input and create a model, which serves as input for the next phase. In this system the data features vary from the set of statistical operations applied to each of the smartphone's inertial sensors data. After creating the model from example inputs with a certain classifier, the model needs to be validated with different sets of annotated data. This process tells us the ratio of correctly and incorrectly classified data points in a confusion matrix. The last phase is to feed non-annotated data to machine learning algorithm in order to be classified. The algorithm will compare the non-annotated data points with the model built in the learning phase and decide on the activity represented by this new data points.

For producing a complete description of the classification function in a learning algorithm, a cluster of candidate applications are consistently provided. Nevertheless, it is predominantly the case where major candidates are inconsequential to the erudition assignment, leading to the problem of over fitting, and this will depreciate the efficiency of the engaged learning algorithm. The indoctrination speed as well as learning veracity may be momentously retrograded by these excessive features. Hence it is of basic concern to exquisite the accordant and significant features in the processing step. This chapter delineates fundamental feature selection issues and current research objectives. Machine learning is an evolving field today due to an increase in big data. It aids to progress observations from a huge quantity of data, which was very ponderous to humans and at times also inaccessible. Machine learning is a specialization of computer, which is an expeditiously trending topic in today's context and is anticipated to bang more in the upcoming years. In a speculation, more discriminating power can be attained by aggregating the size of the feature vector. However, the learning procedure and the model generalization are slow-paced and compromised due to extensively corpulent feature vector. Feature selection is specifically necessary when one is manipulating a vast dataset with dimensions up to thousands.

The prime applications of feature selection are as follows: (i) abbreviating the measurement cost and accommodation requirements (ii) due to the finiteness of training sample sets, coping with the degradation of the classification performance (iii) minimizing exploitation and training time and (iv) promoting data understanding and data conception. It may be visualized that evaluating the significance of distinctive description may not be crucial; nevertheless, the actual remonstrance is to evaluate the consequence of subsets of features.

Data is being created continuously every day throughout the world. According to the Solutions Company CSC, a big data analytic company, the data evolved at the year 2020 could be 44 times larger than the year 2009. Hence, it is essential to recognize data and add observation for the better indulgent of the human world. The quantity of data is so large these days that conventional procedures cannot be used. Evaluating data or constructing extrapolative models by hand is almost infeasible in some consequences and may consume more time and take away productivity [27]. Machine learning produces consistent, imitable results and educates itself from former computation.

There are two types of data used in machine learning. One is labelled data and the other is unlabelled data. In labelled data the attributes are essential and necessary that could imply a tag to the information used in supervised learning, which could be categorical or numerical. To estimate a value of regression, numerical data could be used, and categorical data could be utilized for classifying. Unlabelled data are implemented in unsupervised learning since they have only data points. Due to this implementation the machine can find the structures and patterns present in the data set. Hence these two data types of machine learning can be utilized with supervised as well as unsupervised learning correspondingly. Supervised learning carries out a learning map flanked by a set of variables, say X and Y, which are the Input and Output variables. This mapping is enforced to anticipate the output for invisible data. After gaining the set of data, algorithms simplify the data and articulate the refutable value H for the given sets of data as shown in Fig. 14.8.

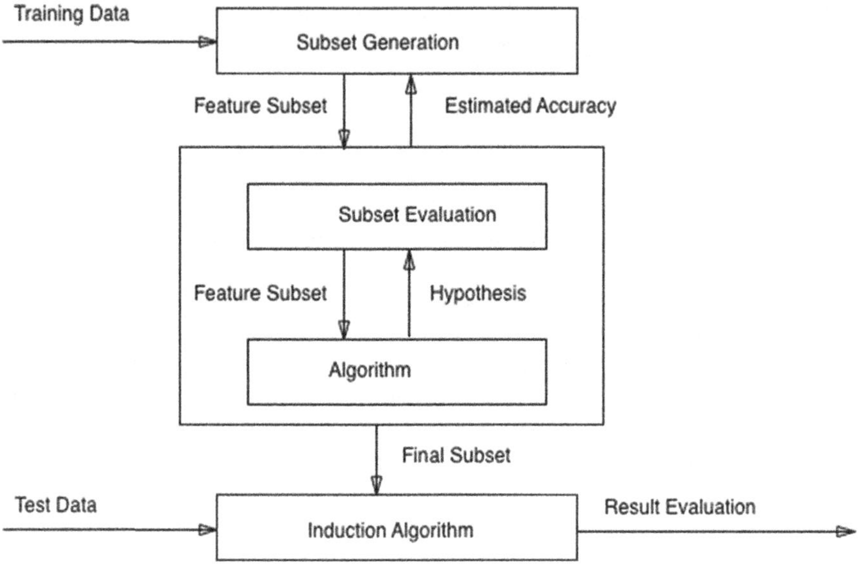

Fig. 14.8 Feature selection process

There are further two types of classifications in supervised learning. They are known as Regression and Classification. A regression could be used within certain variables for a statistical relationship between two or more variables. Classification arises very commonly in day to day life.

It contains partitioning up objects. Each object is accredited to one of the several mutually exhaustive and exclusive kinds known as classes. Each object should be authorized to specifically one class, i.e. more than one object shouldn't be attached to a single class. Unsupervised learning examines how systems could acquire to signify meticulous input patterns in a path that redirects the statistical structure of the general collection of input patterns. A contradiction to supervised learning is that there are no specific objective outputs or environmental estimation accompanying every input.

Feature selection:

This framework comprises two parts: (1) utilization of search engine to obtain the subset of the feature and (2) obtaining the best candidate from the given criteria. The inspection for a variable subset is a NP-hard enigma. Therefore a better solution could not be assured till an exhaustive search is affected in the solution space. The filter method could be the best for the classifier, while a wrapper method utilizes a classifier to calculate feature subsets. This could form the selection scheme, and these methods are plentiful and they could be utilized effectively for the classifier. Filtering methods such as bilateral information, autonomous component analysis, class detaching measure, or variable ranking could be effectively used for classification. In most of the datasets, all the attributes may not help in defining or determining class labels. Enhancing the size of an attribute vector may draw more discriminating power. The excessive attribute vectors affects the learning process, which cause the classifier closely bounded with the training data and compromise model generalization. An attribute congruity can be measured with two strategies: (1) univariate approach and (2) multivariate approach. Univariate strategies are flexible in terms of speed and structure. Most probably inter-relationship and dependencies among the attributes are not counted in univariate, whereas it is well concerned in multivariate search techniques. Multivariate searches also have some restrictions. Initially, they are exposed beyond their limit, mainly in settings. Furthermore, while handling a large attribute arena, it becomes too expensive.

Though ranking of attribute is undefined, its computational and statistical scalability makes it desirable to various variable subset selection strategies. Since it claims only the computation and sorting of n scores it becomes more efficient. The main disadvantage of attribute ranking is that it leads to the assortment of iterative subset. This problem can be resolved by using a smaller subset of equivalent attributes. Attribute ranking can also be achieved by ranking metric and discarding all features that do not reach an enough value.

Due to the high predictability in neural network, it is referred to as feed-forward network. In the current automated industry, we use the backward propagation that comes under the concept of supervised neural networks as shown in Fig. 14.9. The development in this rule has been done that arouses an interest to know more regarding the artificial neural network. In this rule we follow the method of computation efficiency, which could provide changes in the differential function activation units. With this method we would be able to learn the input and output data very effectively. This method of backward propagation was first adopted in the Harvard University. Several procedures have been adopted to improve the backward propagation speed,

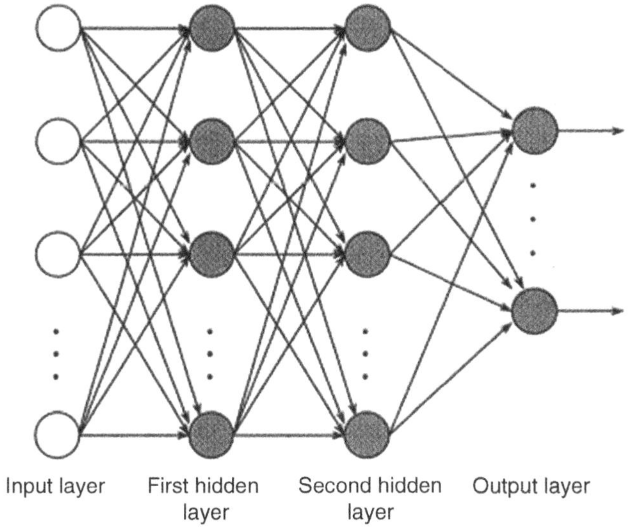

Fig. 14.9 Two-dimensional neural network

tolerance effect, local minima etc. This could be used to solve the diversity problem that could be classifying patterns, function approximation, prediction of time-series as well as developing the non-linear systems.

Machine Learning Influence in Control Systems-Based Industries

Control systems act as the constituent division in this industrial world. The control system is designed to automate the process without the need for human intervention. This could avoid some human errors during the process. This automation process could be greatly supported with the PLCs, which could be used in the process of designing, testing and optimizing the process. The basic advantages of automating a system are increased production, accuracy, safety to the environment and flexibility [28].

There are four main features which make the automation process to be in the lead of the control system engineering:

- Transfer of information from the analogue system into the digital format, which could be supported by the computer-based systems.
- The control methods for the process of automation are carried out with the help of the language-based variable, with the exception of the heavy and tedious mathematical calculations [29].
- Amendment of communication is done by the human intervention with less effort for processing the information [30].
- Utilization of multifaceted automation processed system had increased the productivity in the recent times.

Nowadays, the automatic detection of the technology around the environment is possible that could detect the happenings in real time, which takes all the necessary information to enhance the performance [31]. Meanwhile, when there is less possibility of real-time response, certain malevolent activities, human errors and operational deficiencies occur [32]. Several such IoT-based sensors and machine learning applications are available on the shelf. For example, e-Gauge energy meters connect to the building's electric panel and measure its aggregate electricity consumption every second. These energy meters can upload the measured data to cloud, which stores, analyses and displays the data for the users, thereby helping them monitor and curtail their usage [33]. Nest's smart learning thermostat is an excellent example of Internet-enabled smart home appliance. The Nest thermostat can learn home occupancy patterns and program itself. If the user forgets to turn off heat before leaving, the thermostat will take care of it. The thermostat can be programmed from anywhere over the Internet. The manufacturers claim, by doing all these smart things, that it can cut heating and cooling costs by up to 15%. Besides, there are several other smart home appliances and sensors available in the market. Besides employing Internet-enabled smart appliances, there has been work on programmatically regulating domestic electricity consumption profile.

Conclusion

Initiatives made for the improvement of the quality, estimation of the manufacturing costs, customer requirements and optimization of the process should be best understood by the manufacturing industries to obtain the maximum benefits. Moreover, for complex dimensions and dynamics, support is required to handle them all. Other improvements made in the field of machine learning include knowledge discovery, data mining and artificial intelligence. Still the wide field of machine learning includes various algorithms and methods that are still evolving.

References

1. Alpaydin, E. (2010). *Introduction to machine learning* (2nd ed.). Cambridge, MA: MIT Press.
2. Apte, C., Weiss, S., & Grout, G. (1993). Predicting defects in disk drive manufacturing: A case study in high dimensional classification. In *IEEE Annual Computer Science Conference on Artificial Intelligence in Application* (pp. 212–218). Los Alamitos, CA.
3. Azadeh, A., Saberi, M., Kazem, A., Ebrahimipour, V., Nourmohammadzadeh, A., & Saberi, Z. (2013). A flexible algorithm for fault diagnosis in a centrifugal pump with corrupted data and noise based on ANN and support vector machine with hyper-parameters optimization. *Applied Soft Computing, 13*, 1478–1485. https://doi.org/10.1016/j.asoc.2012.06.020.
4. Çaydaş, U., & Ekici, S. (2010). Support vector machines models for surface roughness prediction in CNC turning of AISI 304 austenitic stainless steel. *Journal of Intelligent Manufacturing, 23*, 639–650. https://doi.org/10.1007/s10845-010-0415-2.
5. Chand, S., & Davis, J. F. (2010, July). What is smart manufacturing? *Time Magazine*.
6. Cherkassky, V., & Ma, Y. (2009). Another look at statistical learning theory and regularization. *Neural Networks, 22*, 958–969. https://doi.org/10.1016/j.neunet.2009.04.005.
7. Chinnam, R. B. (2002). Support vector machines for recognizing shifts in correlated and other manufacturing processes. *International Journal of Production Research, 40*, 4449–4466. https://doi.org/10.1080/00207540210152920.
8. Cohn, D. (2011). Active learning (p. 10). In C. Sammut & G. I. Webb (Eds.), *Encyclopedia of machine learning* (p. 1058). New York, NY: Springer. https://doi.org/10.1007/978-0-387-30164-8.
9. Cook, D. F., Zobel, C. W., & Wolfe, M. L. (2006). Environmental statistical process control using an augmented neural network classification approach. *European Journal of Operational Research, 174*, 1631–1642. https://doi.org/10.1016/j.ejor.2005.04.035.
10. Corne, D., Dhaenens, C., & Jourdan, L. (2012). Synergies between operations research and data mining: The emerging use of multi-objective approaches. *European Journal of Operational Research, 221*, 469–479. https://doi.org/10.1016/j.ejor.2012.03.03.
11. Udendhran, R. (2017). A hybrid approach to enhance data security in cloud storage. In *ICC'17 Proceedings of the Second International Conference on Internet of things and Cloud Computing at Cambridge University, United Kingdom*. March 22–23, 2017, ISBN: 978-1-4503-4774-7. https://doi.org/10.1145/3018
12. Suresh, A., Udendhran, R., Balamurgan, M., et al. (2019). A novel internet of things framework integrated with real time monitoring for intelligent healthcare environment. *Springer-Journal of Medical System, 43*, 165. https://doi.org/10.1007/s10916-019-1302-9.
13. Suresh, A., Udendhran, R., & Balamurgan, M. (2019). Hybridized neural network and decision tree based classifier for prognostic decision making in breast cancers. *Springer - Journal of Soft Computing*. https://doi.org/10.1007/s00500-019-04066-4.

14. Operator Screen (HMI) Design Guidelines. Retrieved from www.hexatec.com
15. Anderson, R. J., & Davies, B. (2010). *Using virtual objects to aid underground storage tank teleoperation*. Albuquerque, NM: Sandia National Labs.
16. Backes, P. G. (2008). *JPL space station telerobotic engineering prototype development: Advanced telerobotics system technology*. Pasadena, CA: Jet Propulsion Lab.
17. Bajcsy, R., Enciso, R., Kamberova, G., Nocera, L., & Sara, R. (1998). *3D Reconstruction of Environments for Virtual Collaboration*. Proc. 4th IEEE Workshop on Applications of Computer Vision, Princeton, NJ.
18. Bourhis, G., & Agostini, Y. (1998). Task sharing and trading for an intelligence powered wheelchair: Problematics. Symposium on Robotics and Cybernetics. In CESA '96 IMACS Multiconference. *Computational Engineering in Systems Applications Journal of Intelligent and Robotic Systems archive,22*(3–4), 269–287. Norwell, MA: Kluwer Academic Publishers. https://doi.org/10.1023/A:1007943113785.
19. Bourhis, G., & Agostini, Y. (2008). The Vahm robotized wheelchair: System architecture and human-machine interaction. *Journal of Intelligent and Robotic Systems: Theory and Applications, 22*(1), 39–50.
20. Bourhis, G., & Agostini, Y. (2013). Man-machine cooperation for the control of an intelligent powered wheelchair. *Journal of Intelligent and Robotic Systems: Theory and Applications, 22*(3–4), 269–287.
21. Brutzman, D. (2011). Virtual world visualization for an autonomous underwater vehicle. Challenges of Our Changing Global Environment. In *Conference Proceedings*. California, USA.
22. Chen, P. C., & Hwang, Y. K. (2006). Generating natural-language like robot motion command set through machine learning. In *Proceedings of IEEE International Conference on Systems, Man and Cybernetics* (Vol. 1, pp. 114–119). Piscataway, NJ: IEEE.
23. Al-Fuqaha, M., Guizani, M., Mohammadi, M., & Aledhari, M. A. (2015). Internet of things: A survey on enabling technologies, protocols, and applications. *IEEE Communication Surveys & Tutorials, 17*(4). Fourth Quarter 2015 (PDF) Internet of things in marketing and retail. Available from: https://www.researchgate.net/publication/312290899_Internet_of_things_in_marketing_and_retail.
24. Andersson, P., & Mattsson, L. (2015). Service innovations enabled by the "Internet of Things". *IMP Journal, 9*(1), 85–106. Friedrich, H., Muench, S., Dillmann, R., Bocionek, S., & Sassin, M. (2009). Robot programming by demonstration (RPD): Supporting the induction by human interaction. *Machine Learning, 23*(2–3):163–189.
25. Graves, A. R. and Czarnecki, C. (2007). Distributed generic control for multiple types of telerobot. In *International Conference on Robotics and Automation*. Detroit, MI, USA.
26. Berndt, H. (2015). *Creating an IoT Ecosystem through scenarios*. Lisbon: IEEE.
27. Caputo, A., Marzi, G., & Pellegrini, M. (2016). The internet of things in manufacturing innovation processes: Development and application of a conceptual framework. *Business Process Management Journal, 22*(2), 383–402.
28. Jagersand, M. and Nelson, R. (2010). Visual space task specification, planning and control. Proceedings of IEEE International Conference on Computer Vision, IEEE, Piscataway, NJ. Pp. 521-526.
29. Krauss, C., Do, X. A., & Huck, N. (2017). Deep neural networks, gradientboosted trees, random forests: Statistical arbitrage on the S&P 500. *European Journal of Operational Research, 259*(2), 689–702.
30. Kshetri, N. (2017). The economics of the Internet of Things in the Global South. *Third World Quarterly, 38*(2), 311–339.
31. Davenport, T. H., & Lucker, J. (2015). Running on data. *Deloitte Review, 16*, 5–15.
32. Sawaragi, T. (2009). Design of human-artifacts symbiotic systems under resource boundedness. *Journal of Japanese Society for Artificial Intelligence, 13*(3), 371–378.
33. Sklyar, V. (2016). Top ten technologies for IoT [Online]. Available at: https://habrahabr.ru/post/308892/

A. Suresh is currently working as the Professor & Head, Department of the Computer Science and Engineering in Nehru Institute of Engineering & Technology, Coimbatore, Tamil Nadu, India. He has nearly two decades of experience in teaching, and his areas of specializations are Data Mining, Artificial Intelligence, Image Processing, Multimedia and System Software. He has published 82 papers in international journals. He has published more than 40 papers in National and International Conferences. He has one patent in the name of Rear view optimised safety helmet. He has published five book chapters for IGI Global Publisher. He has served as a guest editor/reviewer for Springer, Elsevier, and Inderscience journals, etc. He is a member of IEEE, ISTE, IACSIT, IAENG, MCSTA, MCSI, and Global Member of Internet Society (ISOC). He has organized several National Workshops, Conferences and Technical Events. He is regularly invited to deliver lectures in various programmes for imparting skills in research methodology to students and research scholars. He has published four books, namely 'Data structures & Algorithms', 'Computer Programming', 'Problem Solving and Python Programming' and 'Programming in 'C' for DD Publications, Excel Publications and Sri Maruthi Publisher, Chennai, respectively.

R. Udendhran pursued Master of Technology in Computer Science and Engineering from Bharathidasan University. He has presented his research work in the University of Cambridge at the United Kingdom, which is available in ACM digital Library. He has his research publications included in SCI indexed database. Currently, he is pursuing PhD in deep learning in Bharathidasan University.

M. Balamurugan is the Professor and Head of the Department of Computer Science and Engineering at Bharathidasan University. He has completed his PhD and guided 10 research scholars. His area of research includes cognitive science and data science. He has completed a UGC-funded research project, has two patents and has published 30 papers in reputed journals.

Chapter 15
Machine Intelligence and Automation: Deep Learning Concepts Aiding Industrial Applications

S. Sree Dharinya and E. P. Ephzibah

Introduction

Nowadays artificial intelligence with deep learning is a thriving area in which more research and better practical tools and application are built. Earlier, the complex problems were approached with solutions based on mathematical concepts and now that they are looked at with a more intuitive approach a simpler and practical way is adopted. To bring about solutions to be simpler and also to avoid complicated concepts, computers build on concepts with many layers called deep learning with the use of artificial intelligence. The most difficult part is the hard-coded technology that the AI uses for the patterns. The solutions to such problems which are more subjective with the knowledge of real-world technology have been the evolution of machine learning [1].

Deep learning is used in various applications such as image analysis and text identification. It uses supervised and unsupervised learning to represent data in a hierarchical format. Although big data has great opportunities, it had some constraints on data mining and information processes due to the characteristics of its volume [2]. The growth of sensor networks and communications has posed a great insight into deep learning for the usage and classification and applications of machine learning. The digital transformation of industry which includes the smart systems is towards the Industry 4.0. Smart systems and factories with cyber physical system are growing towards Industry 4.0. Deep learning techniques are capable of identifying the complex signal which are sent and received [3]. Digital manufacturing, smart factories and applications with speech and text recognition are some of the perspectives that drive the deep learning techniques along with IoT to develop and get into Industry 4.0.

S. Sree Dharinya (✉) · E. P. Ephzibah
School of Information Technology and Engineering, Vellore Institute of Technology, Vellore, Tamilnadu, India
e-mail: ssreedharinya@vit.ac.in

© Springer Nature Switzerland AG 2020

G. R. Kanagachidambaresan et al. (eds.), *Internet of Things for Industry 4.0*, EAI/Springer Innovations in Communication and Computing,
https://doi.org/10.1007/978-3-030-32530-5_15

The data is stored in large volume and thus the big data has an explosive amount of data. This volume of large data is both in structured and unstructured forms. Nearly three-fourths of the data are unstructured, which is generally the data from internet and mobile applications [4]. Beyond the volume of the data, the large exactness aids in inaccurate objects and blaring data and replicated data objects. Analyzing real-time data, like an online data becomes very critical in case of big data. Handling the massive data becomes difficult with applications in health care or industry, as the data to be extracted is to be intuitive and precise. The major constraint in data mining is the large data and is very difficult to analyse the large variation of data. Deep learning plays a vital role in harvesting the targeted data from the huge volume of data and also from difficult systems [5]. Deep learning takes back to the early techniques where the data is pre-trained and perfected for effective data mining.

Deep Learning Models

The deep learning concept arises when a set of data is to be retrieved from a complex set of data or environment. Figure 15.1 represents a deep learning model. According to the concept of artificial intelligence, it is difficult for the computer to understand raw data given by sensors which is similar to the image given in the

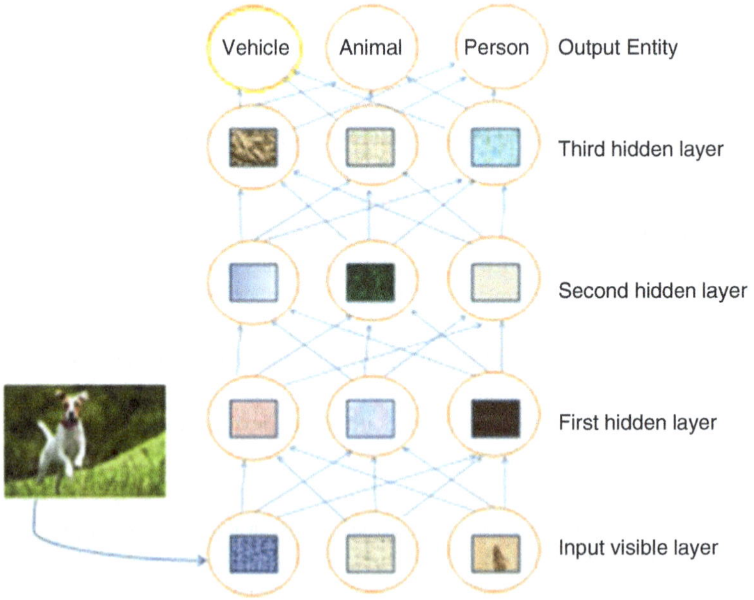

Fig. 15.1 Deep learning model

figure. The way in which the pixels of the picture are mapped is quite complex and it seems next to impossible once we think of solving it directly [6]. Using the technique of deep learning it is possible to identify the object with a sequence of nested mapping, and the layers of each mapping depict the different attributes of the object. Here the input data is shown in the observable layer and is also termed likewise as it has all the features of the object which is to be identified. Then consecutively the set of hidden layers are available, which reveal the abstract properties of the image. The layers are termed as hidden layers since the data to be extracted is not based on the values but on the underlying concept of mapping the object to be established. The pictures here are conceptions of the type of data which is hidden. The first layer compares with the limits related to the brightness of the picture and the second layer identifies the angles. They are collectively searched as a data to view the picture. The next layer finds the precise objects related to angles of the picture. The final layer is the one which describes the parts of the object which it has to identify in the picture [7]. Deep learning does not invoke the factors of each layer but it helps to run a program with sensible input [8].

Deep learning thereby helps to organize the data rather than the exact content of the data. The data is not visible at the moment the object is identified. After layer of identification and identification of edges and contours, the data is visualized as an object in the last layer. Such techniques evolve the types of deep learning with various approaches based on the output parameters. For a large data set where the identification of the object is complex, deep learning uses the method of identifying the concept as an outcome. Such techniques aid with the machine learning algorithms and also they are used in various sectors as industries, smart home and smart city developments with the help of IoT. The category of learning from a complex set of data is done using deep learning concepts.

Industries make use of the data available in large volume and indulge in data-driven applications for their benefits like easy access to information, knowledge to upgrade and manufacture the products in demand, increase the income and so on. They like to optimise the manufacturing operations using machine learning tools and techniques. A company that manufactures steel pipes for oil and natural gas industries makes use of the data science team to combine artificial intelligence in the area of automation solutions. Different data sources are integrated and the system becomes flexible with the help of machine learning algorithms that are far easier and faster [9]. Deep learning helps to test different models and chooses the best model in limited time; thus, predictions are done in seconds. Result analyses and performance reports are efficiently done faster so as to obtain customer satisfaction. In a real-time system, it allows to control energy consumption, produce better products, save cost, deliver to the customers on time. Thus, they provide numerous ways for the industries to be a leader among their competitors with traditional approaches [10].

Deep learning neural networks are one of the principal machine learning techniques in the recent years. Many of the well-known and commonly used applications that use deep learning neural networks are speech recognition, face recognition, autonomous or self-driving cars, image tracking and computer vision [11].

Deep learning is widely used in industries, as it takes very complex concepts and composes them out of smaller and smaller concepts to capture something very complicated [12].

According to the medical industry, drug discovery is a major area where deep learning can be applied to find a fit in terms of the drug with the proper disease [13]. It is a flourishing field in all the countries of the world, as it captures the special benefits of the drugs and their demand among the people in curing the diseases. Using artificial intelligence and deep learning, it is more practical to cure any kind of disease.

Deep learning uses large volume of data especially in medical field where a tremendous increase in the volume of data over a period of time is possible. For example, if the heart rate of a patient is to be monitored for every second and the data is to be recorded, imagine the volume of data for a period of 10 days or 1 month [14]. Deep learning is the state-of-the-art machine learning method especially for medical imaging and image analysis.

Deep learning and high-performance computing are enabling the new industrial revolution –an era of smart creation and streamlined production [15]. Manufacturers have been quick to adopt deep learning, artificial intelligence and machine learning using their ability to improve processes across supply chains, asset management, predictive maintenance pricing and more with such podium-packed automation, cloud technology and streamlined IT (Information Technology) are more important than ever before. With deep learning, manufacturers are already exploring and simulating designs in virtual environment. Intelligently managing logistics and printing 3D prototypes their optimizing costs at each step by speeding up design and testing and keeping machine to be running to minimize down time. Soon advances in deep learning driven by powerful GPU accelerators will make virtual products development even more accurate by simulating models with unprecedented accuracy. Environments in various fields make sophisticated digital twining a reality for manufacturers. Minimizing expensive designing process at production stage is again challenging. Robotic machinery, however, contributes to cost-saving by identifying downtime opportunities and shutting itself down. Service operations are revolutionized by predictive maintenance that optimizes the equipment operational life time and stops failures before they even occur by embedding and managing AI in new infrastructure. Many industries work for this kind of operations. Especially companies like HPE allow AI to work for the business and for IT. The data scientist, innovators and big data architects and those leading product engineering companies support digital evolution within manufacturing.

Artificial intelligence and high-performance computing are changing the landscape in both e-commerce and physical retail shops. Retailers are using artificial intelligence, machine learning and deep learning to attract, to engage and to maintain relationships with customers improving both efficiency and loyalty. In such a highly competitive environment, automation cloud technology and streamlined IT are important than ever before. Online retailers rely on AI recommendation engines to thoroughly analyse the activities of the potential customers and drive traffic to their sites, and in physical stores AI is planning and guiding shoppers' experience with smart layouts and inventories checkout free technology and video analytics. With deep learning, retailers would soon be exploiting the potential of personalized digital assistance to make recommendations. Visual search in stores define the right item and even virtual reality to fit in for a quicker buying process. The future potential of AI in retail is huge with only minimum percent of companies confident in responding to far changing demands. Embedding and managing AI in the infrastructure makes the business faster and more productive with lots of profit [16–18].

AI and deep learning also have a very great impact on the dairy farms and milk products industries. Cows in the form require more attention and technology help the owners to make decisions for the better growth in cows and for good income. Artificial intelligence acts as a great tool to get more insights. It helps to keep track of the cows and makes the farmers life easier. Sensors are connected to the neck of the cows [19]. By analysing the movement of the sensors, the machine learning algorithm figures out the movement of the cow and how it is doing. It is also possible to distinguish the behaviours of the cow like eating, drinking, standing, lying and walking. The software identifies the disease the cow is prone to and can warn the farmer. Using tensor flow and open-source tools it learns the behaviour of the animals and it gets better over time. It improves the efficiency of milk production, animal welfare to keep the cows healthy.

In any business anticipating and planning what is next is crucial for success. When caring for someone's health, predicting what is ahead has the potential to save lives. There are so many AI tools and techniques along with machine learning to achieve the task. An initiative that will make machine learning and predictive analytics widely available across all of health care is on the move. Today it is observed that only the largest health systems have the resources to build and play the predictive analytics [20]. Demand for data scientist having the right people on board is either cost prohibitive and sustainable or possible. These lifesaving technologies should be made available everywhere. Machine learning models are designed and built into each product that is delivered. Caregivers can now easily make predictive analytics routine and actionable. Imagine how valuable access to data-driven machine learning could be for caregivers like data mutineers. Utilizing machine learning, data mutineers can discover patterns in their own patient's data and they can identify the best interventions to reduce the patient's length of stay and predict the next deadly infection. Based on patients' specific risk factors the deep learning and AI techniques could even prevent the next onset of diabetes; these benefits could be made available to many in health care who are not just as customers. There are companies that take this technology and learning in creating healthcare systems using deep learning and artificial intelligence [21–24].

There are lots of applications that make use of the deep learning algorithm to find solutions to the real-time complicated and time-consuming problems. Figure 15.2 given below depicts a few of them.

Deep learning has become a well-known ignition technology behind many devices such as amazon alexa, siri on i-phone and many other voice-enabled devices. At the core of this learning is the use of artificial neural networks which are computer programs that enable machines to learn. They are inspired by some of the

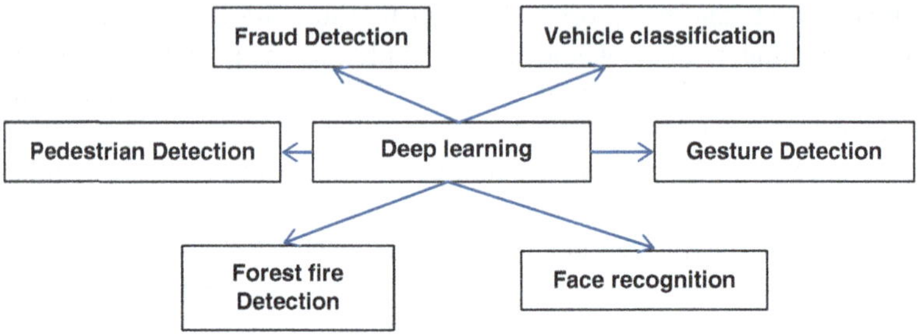

Fig. 15.2 Applications of deep learning

computation that goes on in our brain, real neural networks. Consider a situation of building a machine that can read and write information. At the core of artificial neural networks are the artificial neurons that are connected. The neuron can be connected to pixels of the image. The job of the artificial neuron is to detect patterns from its incoming connections. Like human brain that has many neurons, artificial neural networks have many neurons, each one performing a different operation on technically different types of data. In deep artificial neural networks there are multiple layers, and this is the core idea behind deep learning.

Types of Deep Learning Models

The various types of deep learning techniques are classified under categories such as supervised, unsupervised and semi-supervised. Supervised learning is grouped as regression and classification problems. The classification problem is based on the category which is attribute-based and the regression is based on the original value like predicting the time series.

Unsupervised learning is based on the input of the data and has no relevant output parameters. The dissemination of data is more focused as there are no exact answers and no specific mentors. New algorithms are devised and are classified and identified for clustering and association problems. Clustering helps to identify the group behavior, and association helps to discover large data. They tend to solve the learning problems associated with unsupervised learning algorithms. Industry uses communication systems to implement the artificial intelligence concepts. The traditional type of communication system is shown in Fig. 15.3. They work in blocks and in a linear format where there are responsibilities assigned to each block. This type of communication becomes more difficult and more prediction-based. The robustness of the system is the thought process when the blocks undergo certain flaws. A simulation-based communication system can be correctly interpreted using machine learning algorithms.

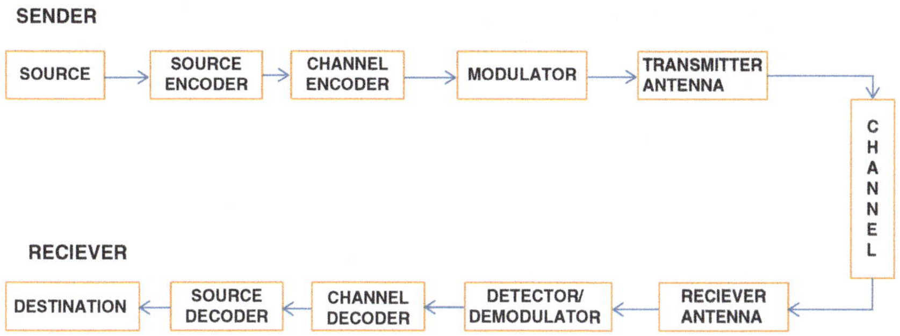

Fig. 15.3 Traditional communication system

Applications of Deep Learning in Industries and Smart Applications

Industries use different artificial intelligence techniques. In warehouses one such air abnormality detection in industries is identified through deep learning techniques. Warehouses are an integral component of the transport and logistics industry, and tremendous amounts of operational efficiencies can be unlocked by introduction of intelligent IoT solutions for Warehouse Management. One such solution is real-time hazardous chemical or gas detection that ensures safety of warehouse by detecting hazardous, inflammable chemicals/gases that could get leaked from items stored in a warehouse. The deep learning method called HMT algorithm for detection of hazardous gas and chemicals in warehouse is used widely.

Toxic gases have huge environmental effects. People are suffering several diseases due to them. Proper detection of toxic gas level is important for us. Toxic gases have huge environmental effects. Sometimes when highly inflammable and hazardous gas/chemicals leak it leads major accidents, life loss, and severe health issues. The protected transportation of these chemicals and gases are to be concentrated in high range, because only during transportation large quantity of them are transferred. Minute leakage also may lead to major accidents such as life and property loss, air pollution and affects health from outer skin to micro DNA in living things. These effects last for generations to generations.

The existing systems use different set of sensors like MQ2 and MQ7; MQ9 and MQ4, MQ2 and MQ7 gas sensors for detection which do not cover some hazardous gas. These systems use machine learning algorithm which was trained by some predefined set of training data set by rule-based data analysis for finding anomaly and release of toxic gases. It also uses "Deep Belief Network" (DBN) as an unsupervised learning algorithm to find the generic features of the input. That output is fed into SVM trained in a way to detect whether there is abnormality present in the data with single class output.

Prediction is processed online using the algorithm, as it identifies state xt as usual or anomalous earlier getting the following $xt + 1$. This procedure studies continuously without a requirement to store the entire stream. This procedure runs in an unsupervised, automatic fashion without data tags or manual parameter alteration and adapts to vibrant surroundings and idea drift, as the underlying statistics of the data stream is often non-stationary. It also makes anomaly detections as early as possible and minimizes false positives and false negatives which are true for batch scenarios as well.

Environmental Gas Sensors

Detecting principles of solid-state smoke sensors and gas sensors which are used to detect the air pollution level must be able to work in any harmful condition irrespective of thermal or chemical attack.

Types of sensors are Semi-conductor and capacitor-type gas sensor, Non-Dispersive Infrared Method using Pyroelectric Infrared Sensors, Solid-Electrolyte. All other techniques which involve detection finished from outdoor the pipeline by way of visual observation or portable detectors are capable of discover very small leaks and the leak place, however the detection time is very long. Although the charges involved for detection seems to be high,the precision obtained is remarkably accountable. This downside is vulnerable to disappear for a number of these techniques due to forthcoming technological improvements [25].

It gives an overview of the pollution caused to the environment. The solid state sensors which may cause hazardous environmental pollution are taken into consideration and conventional instruments are used which in turn is costly and time consuming. Addition to this as we are living in a rapidly developing environment rapid development in the gas sensors are important to live a safe and comfortable life.

Unsupervised real-time abnormality detection for running data explains the problems in detecting anomaly in streaming data and proposed a machine learning algorithm to deal with data. Normally "Machine Learning" and "Deep Learning" models are trained in batches using data which are collected for that particular model. But in case of streaming data its different and we should use real time data's to detect anomaly in our data. For this they proposed Hierarchical Temporal Memory (HTM) and they also present their results using Numenta Anomaly Benchmark (NAB), a benchmark containing real-world data streams with labelled anomalies. The algorithm learns continuously and does not require the data to be stored like in static batch anomaly detection. Even "Time Series Analysis" which is a highly researched area in data science is not applicable for real-time streaming data. Usually HTM learns continuously and the model's spatio temporal characteristics analyses the input and does not detect anomalies. The computations on the output of HTM, provides a better prediction error and there is a likelihood that the system is in an anomalous state. The threshold on this likelihood is used to detect whether an anomaly is found. The number of sensor data and flow of streams of data are increasing and anomaly detection in real time is getting more important and unavoidable. The algorithm framed is capable of detecting spatial and temporal anomalies in predictable and noisy domains.

High-dimensional and Large-scale Anomaly Detection Using a Linear One-class SVM with Deep Learning [26, 27].

Anomaly detection in low-dimensional data is fairly straightforward and easy compared to high-dimensional data. In high-dimensional data, all the data won't be relevant to anomaly and this will conceal the presence of the anomaly present in the data, and this is commonly known as "Curse of Dimensionality" in data science communities. To handle this problem, this chapter proposed a hybrid algorithm which is a combination of unsupervised feature extractor and an anomaly detector. First the most relevant features for anomaly are extracted from the given high-dimensional data and then they are fed into an anomaly detector to find the presence of anomaly in the data. The feature extractor used is "Deep Belief Network" (DBN), which is an unsupervised learning algorithm to find the generic features of the input. That output is fed into SVM trained in a way to detect whether there is anomaly present in the data with single class output. This hybrid algorithm after implementation gave good results with better accuracy of anomaly detection, as it addressed the complexity and scalability issues of 1SVM, especially in large-scale datasets. DBN is used as a dimensionality reduction algorithm and helped solve the problem of scalability of 1SVM model.

Hierarchical Temporal Memory is a machine learning algorithm first proposed in 2005 by a Brain Scientist called Jeff Hawkins. This model is inspired by neocortex which is a part of biological brain. It is a part of the cerebral cortex concerned with sight and hearing in mammals, regarded as the most recently evolved part of the cortex.

The HMT algorithm concept is given such that, in brain, the neocortex cells are very similar to one other, and they do not have any relation to the sensory input organs such as eyes and ears. But the same neocortex handles different inputs and trains itself to gain useful information and predict using this input. This is why HMT algorithms can be used for wide range of applications and not restricted to one particular set of problems. HMT algorithms show promise in building intelligent systems. It can be used to solve problems which are easy for humans but difficult for machines (Vision, Audio). Hierarchical Temporal Memory consists of nodes in a hierarchical tree structure. All these nodes are using same algorithm called as Naïve Bayes. These nodes form the networks in Hierarchical structure with many inputs and few outputs. HMT uses the spatial-temporal concept. This concept acts as the basic and absolutely necessary component of human intelligence. Temporal means with relation to time. In a sequence of inputs the time input acts as key to identify the dynamic changes in the input. Spatial means the space. The static position of the inputs handling both space and time simultaneously might not be easy and so hierarchical structure is used. HMT is different from other machine learning algorithms as it is completely self-learning.

The deep learning techniques use the application IoT for smart applications. As given by Sunil Raj Thota et al. [20], the devices interconnected and controlled through internet are called Internet of Things [3]. There are many methods of solid waste collection and management. Something like using wireless sensor networks,

IoT, Microcontroller-enabled senor networks and many more. Among all the IoT-based sensor network with sensors working in the real time is understood to be the best and can grab the current status of the bin and helps in communicating it via some Wi-Fi module. Also real-time analytics and route optimization algorithms have added up a lot more advantage for the methodology and provide a solution which is more accurate, optimistic and efficient when compared to others. The smart dust bin is connected to the internet to provide real-time information on waste management necessary to avoid diseases by managing smart dustbins. The system can monitor the level of trash can and shows the status of the dust bin level through the mobile app from anywhere through Internet. It is useful and can be fixed in the trash cans at public places, smart cities and other places. For the implementation of smart dustbins, microcontrollers and sensors are used to reduce the cost and to make it an efficient device. To monitor the trash in the dust bin sensors are attached to trash can to detect the level. The system consists of sensor circuitry used for monitoring the smart trash can. The project module is divided into two parts: sensor section and Node MCU Wi-Fi module section. Sensors are attached to the dust bin. Sensors are used to detect the level in the dust bin. The sensor senses the content of the dust bin and sends the signals or the data to the Arduino. Wi-Fi Module helps to send the information of the trash to the receiver side through internet [28].

The waste management in industry is also managed as given by Mohan et al. [29] as the major challenge faced by urban India in current context of urbanization is managing solid waste. The waste generation in 2025 is expected to be double the amount of waste generated in that of 2012. The complexity of solid waste management in India is added by unavailability of infrastructure and local municipal bodies. The customized model emphasizes on door to door collection of waste for 6 days in a week. The waste collector is equipped with hand-pulled carts with two separate bags for recyclable and non-recyclable waste. This model provides daily services around 250 households. This model works as decentralized waste management system. The collected waste is transported by waste collectors for secondary segregation. After that the waste is separated from the waste to make compost in non-biodegradable. Sustainability, sale of recyclable waste and compost are the major sources of revenue. Of the total revenue generated in past 3 years, around 48% comes from user charges. It is a positive indicator for the sustainability of the model.

References

1. Unnikrishnan, S., Razil, M., Benny, J., Varghese, S., & Hari, C. V. (2017). LPG monitoring and leakage detection system. In *2017 International Conference on Wireless Communications, Signal Processing and Networking (WiSPNET)*, Chennai (pp. 1990–1993). IEEE.
2. Shi, L., Wang, X., Zhang, T., Hu, C., Luo, K., & Bai, B. (2016, August). Hazardous gas detection four-rotor UAV system development. In *Mechatronics and Automation (ICMA), 2016 IEEE International Conference on* (pp. 2461–2465). IEEE.

3. Li, S., Wen, Y., Li, P., Yang, J., & Yang, L. (2012, October). Leak detection and location for gas pipelines using acoustic emission sensors. In *Ultrasonics Symposium (IUS), 2012 IEEE International* (pp. 957–960). IEEE.
4. Ajith, S., Harivishnu, B., Vinesh, T. K., Sooraj, S., & Prasad, G.. (2017). Automated gas pollution detection system. In *2017 Second International Conference for Convergence in Technology (I2CT)*, Mumbai (pp. 483–486). IEEE.
5. Wang, C., Olson, M., Sherman, B., Dorjkhand, N., Mehr, J., & Singh, S. (2018, October). Reliable leak detection in pipelines using integrated DdTS temperature and DAS acoustic fiber-optic sensor. In *2018 International Carnahan Conference on Security Technology (ICCST)* (pp. 1–5). IEEE.
6. Spachos, P., Song, L., & Hatzinakos, D. (2014, January). Gas leak detection and localization system through Wireless Sensor Networks. In *Consumer Communications and Networking Conference (CCNC), 2014 IEEE 11th* (Vol. 1130, No. 1131, pp. 10–13). IEEE.
7. Shu, L., Chen, Y., Sun, Z., & Tong, F. (2017). Detecting the dangerous area of toxic gases with wireless sensor networks. IEEE Transactions on Emerging Topics in Computing. http://ieeexplore.ieee.org/stamp/stamp.jsp?tp=&arnumber=7917279&isnumber=6558478.
8. Varma, A., Prabhakar, S., & Jayavel, K. (2017, February). Gas leakage detection and smart alerting and prediction using IoT. In *Computing and Communications Technologies (ICCCT), 2017 2nd International Conference on* (pp. 327–333). IEEE.
9. Alotaibi, M., Lo'ai, A. T., & Jararweh, Y. (2016, November). Integrated sensors system based on IoT and mobile cloud computing. In *Computer Systems and Applications (AICCSA), 2016 IEEE/ACS 13th International Conference of* (pp. 1–5). IEEE.
10. Mhatre, L., & Rai, N. (2017, February). Integration between wireless sensor and cloud. In *I-SMAC (IoT in Social, Mobile, Analytics and Cloud) (I-SMAC), 2017 International Conference on* (pp. 779–782). IEEE.
11. Bokefode, J. D., Bhise, A. S., Satarkar, P. A., & Modani, D. G. (2016). Developing a secure cloud storage system for storing IoT data by applying role based encryption. *Procedia Computer Science, 89*, 43–50.
12. Ahmad, S., Lavin, A., Purdy, S., & Agha, Z. (2017). Unsupervised real-time anomaly detection for streaming data. *Neurocomputing, 262*, 134–147.
13. Erfani, S. M., Rajasegarar, S., Karunasekera, S., & Leckie, C. (2016). High-dimensional and large- scale anomaly detection using a linear one-class SVM with deep learning. *Pattern Recognition, 58*, 121–134.
14. Chen, X., Wang, W., & Li, W. (2012, June). An overview of hierarchical temporal memory: A new neocortex algorithm. In *Modelling, Identification & Control (ICMIC), 2012 Proceedings of International Conference on* (pp. 1004–1010). IEE.
15. Brugo, T., Palazzetti, R., Ciric-Kostic, S., Yan, X. T., Minak, G., & Zucchelli, A. (2016). Fracture mechanics of laser sintered cracked polyamide for a new method to induce cracks by additive manufacturing. *Polymer Testing, 50*, 301–308. https://doi.org/10.1016/j.polymertesting.2016.01.024.
16. Rai, R. K., Bhattarai, D., & Neupane, S. (2019). Designing solid waste collection strategy in small municipalities of developing countries using choice experiment. Journal of Urban Management. https://doi.org/10.1016/j.jum.2018.12.008.
17. Lee, D. D., & Lee, D. S. (2001). Environmental gas sensors. *IEEE Sensors Journal, 1*(3), 214–224.
18. Ramírez, C., & Gonzalez, E. (2019). Methodological proposal for the interinstitutional management of wastes in health care centers in Uruguay. *MethodsX, 6*, 71–81.
19. Yusof, N. M., Zulkfli, M. F., Mohd Yusof, N. Y. A., & Azman, A. A. (2018). Smart waste bin with real time monitoring system. *International Journal of Engineering & Technology, 7*(2), 29.
20. Thota, S. R., Neelima, S., Lalitha Pruthvi, K. V. N., Mouika, K., Pravallika, M., & Sowmya, N. (2018). Smart trash can monitoring system using IoT – Creating solutions for smart cities. *Waste Management, 29*, 2625–2643.

21. Selvaraj, K., & Chakrapani, A. (2017). Smart dustbin monitoring system using LAN Server and Arduino. *International Journal of Advances in Computer and Electronics Engineering, 2*(4), 20–23.
22. Jino Ramson, S. R., & Jackuline Moni, D. (2016). Wireless sensor networks based smart bin. *Computers & Electronics Engineering, 64*, 337–353.
23. Nelles, M., Grünes, J., & Morscheck, G. (2016). Waste management in Germany – Development to a sustainable circular economy? *Procedia Environmental Sciences, 35*, 6–14.
24. Ghosh, S. K. (2016). Swachhaa Bharat Mission (SBM) – A paradigm shift in waste management and cleanliness in India. *Procedia Environmental Sciences, 35*, 15–27.
25. Capel, A. J., Edmondson, S., Christie, S. D., Goodridge, R. D., Bibb, R. J., & Thurstans, M. (2013). Design and additive manufacture for flow chemistry. *Lab on a Chip, 13*(23), 4583–4590.
26. Choi, S., Jun, C., Zhao, W. B., & Noh, S. D. (2015a). Digital manufacturing in smart manufacturing systems: Contribution, barriers, and future directions. In *Advances in production management systems: Innovative production management towards sustainable growth* (Vol. 460, pp. 21–29).
27. Choi, S., Kim, B., & Noh, S. (2015b). A diagnosis and evaluation method for strategic planning and systematic design of a virtual factory in smart manufacturing systems. *International Journal of Precision Engineering and Manufacturing, 16*(6), 1107–1115. https://doi.org/10.1007/s12541-015-0143-9.
28. Samir, K. C. (2014). An age distribution story: Reading into the population pyramid of Nepal. *Asian Population Studies, 10*(2), 121–124.
29. Mohan, G., Sinha, U. K., & Lal, M. (2016). Managing of solid waste through public private partnership model. *Procedia Environmental Sciences, 35*, 158168.

Index

© Springer Nature Switzerland AG 2020

G. R. Kanagachidambaresan et al. (eds.), *Internet of Things for Industry 4.0*, EAI/Springer Innovations in Communication and Computing,
https://doi.org/10.1007/978-3-030-32530-5

The manufacturer's authorised representative in the EU is Springer
Nature Customer Service Centre GmbH, Europaplatz 3, 69115 Heidelberg,
Germany. If you have any concerns regarding our products, please
contact ProductSafety@springernature.com

Printed and bound by CPI Group (UK) Ltd, Croydon, CR0 4YY
30/09/2026
02225495-0008

The manufacturer's authorised representative in the EU is Springer Nature Customer Service Centre GmbH, Europaplatz 3, 69115 Heidelberg, Germany. If you have any concerns regarding our products, please contact ProductSafety@springernature.com

Printed and bound by CPI Group (UK) Ltd, Croydon, CR0 4YY
30/09/2026
02225495-0008